Satellite Spotting Handbook

For the beginner

Peter Bassett F.R.A.S.

I0485876

An inter-active book via our website.

Astronomy Roadshow Publishing
Further paperback copies, signed copies e-books with direct
internet links can be purchased from…

www.outerspacebooks.com

**All of the links in this book plus many more are offered as a
short cut via the satellite spotting website. Just search the
relevant page.**

*Credits to the images have been given wherever possible. Around half
were taken by the author. This book has been in the making for forty
years. I thought it was about time to release it.*

Cover picture by ESA.

Contents

Most Satellite projects have a designated patch that represents the mission. Once a particular satellite is witnessed, purchase the corresponding patch and wear it with pride.

Chapter 1   Introduction

Satellite Spotting is a space age version of Train or Plane Spotting. It is possible to watch satellites passing over your hometown night after night. They look like moving stars of varying brightness. Dozens of satellites can be seen every clear night passing over silently and majestically. We may be completely in the dark down here, but hundreds of miles up, the Sun still illuminate the satellites. Anything larger than a car that is less than 600 miles up can be seen without any binoculars or telescopes. This hobby can be completely free – apart from the few pennies for this book.

This book is written for the complete beginner to an experienced observer; from a youngster at a school with no money looking a for a potential hobby or career to someone aged 90 who want to sit in a deck chair at night and watch these things go silently. A hedgehog may snuffle on the ground wondering what you are looking at (I have an e-book / paperback on Amazon about those too – Attracting and Looking After Hedgehogs). This is a *handbook* for quick reference and not designed as a detailed study that will take hundreds of hours to read. This is designed purely as a reference guide to satellite spotting and satellite operations for perhaps those that may be considering a career in this field or wish to learn a little more about how they work.

Please refer to <u>www.outerspacebooks.com</u> for clips & updates. All links in this book is available via the website. Just look up the corresponding chapter.

The author with his dedicated satellite recording cameras. These are not essential pieces of equipment; your eyes will do.

Chapter 2   A Free Hobby That Can Change The World

If you're on a budget, this hobby has a great advantage; it does not have to cost anything. If you are already on the internet, that is all that is required. If not, ask someone who is for a favour. From this point, anyone can print out a list of satellites, available online for any night, and simply go outside and see them pass over. Knowing which is which and learning their role in our lives makes these amazing machines feel closer.

Any youngster at school – even without any money coming their way – can take up this space age interest. They can show off to their friends that they have seen Envisat or the International Space Station pass over their garden the previous night as they waved to the astronauts. This would not be some wild fantasy but a true reality.

In the UK alone, some 80,000 people are employed designing, building and operating satellites, an industry that is worth £9.3 billion ($16 billion) a year as of 2018. Growth took it to £19 billion ($28 billion) in 2020, and it is predicted to reach £40 billion ($60 billion) by 2030. An extra 100,000 related jobs will be created. Perhaps some of the youngsters at school today may gain a spark of enthusiasm in this field from satellite spotting and consider something related to it as a career. Britain produces statistically some of the most reliable satellites in the world.

Here are some such places of employment in the UK...

Airbus Defence and Space (formally known as Astrium / British Aerospace / Matra Marconi) Stevenage, Hertfordshire & Portsmouth, Hampshire).
Surrey Satellite Technology Ltd (Guildford).
Mullard Space Science Laboratory (Guildford).

Clyde Space, Scotland.
Stevenson Astrosat, Edinburgh, Scotland
Tessella, Abingdon, Oxfordshire and Stevenage, Hertfordshire.
Canterbury University.
Leicester University.
Teledyne Technologies (www.teledyne.com)

A short listing of US satellite builders / contractors…
US Air Force
US Navy
US Army
Bell Aerospace
Boeing
General Dynamics
General Electric
Lockheed Martin
Northrop Grumman
TRW
Spacequest
Orbital Sciences Corporation
Los Alamos
Ames Research Center
Jet Propulsion Lab
Marshall Spaceflight Center

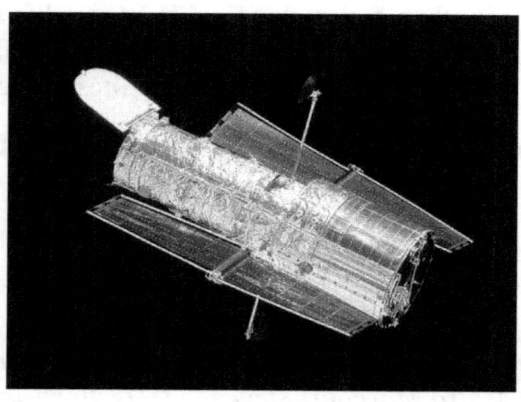

British Aerospace (now Airbus – Defence & Space) built the original Solar Panels for the Hubble Space Telescope in Bristol. Teledyne Technologies built the camera sensors.

The space industry was the only area that expanded from 2008 – 2013 during the banking crisis, needing very little help from government. A strong 7.5% growth occurred throughout the recent recession, and as of 2019 it reached 12%. This clearly demonstrates the solidity of such a market. All other areas such as manufacturing, finance, tourism, construction, public services etc. all suffered cutbacks, wage freezes and job losses. This area of space research, communications and space-based observations is in great demand. Such new data improves weather forecasting, land management, disaster monitoring, inspires new inventions and general scientific discoveries. These in turn produce new industries and create jobs for the future.

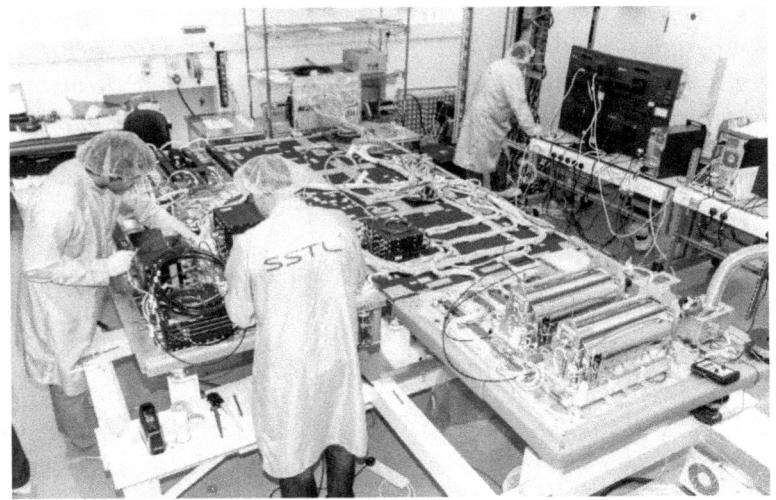

Technicians working at Surrey Satellite Technologies LTD, Guildford, Surrey, UK.

The lure of working in the glamorous space industry has helped a company called Tessella in Oxfordshire to recruit some of the best brains in the business. One of the company directors, Mr Whittle, once famously said…

"From when we are kids, it's all about space and dinosaurs. Would you rather work in a bank or somewhere you can make a real contribution to your planet?" (His words not mine; I do not want my bank account frozen).

A career in the space industry (or banking) can involve making a good living, but making a difference to the world and potentially changing future history in a positive way is priceless. Work is not just about making money. I believe making an impact and changing the world for the better is far more important.

If you do not happen to live in the USA, please do not feel that it will hold you back. The USA does indeed produce some amazing technology and is well known as a spacefaring nation. Remember the following simple points:

*The laws of motion that control orbits were discovered by Sir Isaac Newton (**British**).*

*The original mathematical principles to get into orbit and maintain or change it were discovered by Konstantin Tsiolkovsky, (**Russian**), in case you did not guess already.*

*The first person to write about communication satellites was the author Arthur C. Clarke, (**British**).*

*The first rockets into space were developed by Wernher Von Braun, (**German**).*

*The very first satellite was Sputnik 1, (**Russian**).*

*The practical method that was chosen to reach the Moon was originally developed by the British Interplanetary Society, (**British**) – not NASA.*

*The most distant landing to ever take place, Saturn's moon Titan was achieved by the **European Space Agency, (ESA),** which includes **France, Germany, Belgium, UK,** etc.*

*The only landing on a Comet so far, (Philae Lander on Comet 67P/Churyumov-Gerasimenko), was accomplished by **ESA**.*

This could go on page after page. Careers in satellite and spacecraft design, construction, launch, communications, data handling, tracking etc. are worldwide. Over five million jobs are directly related to the exciting field that is changing the way we live and look at our fragile world.

Chapter 3   If you want to spend...

As mentioned in the previous chapter, Satellite Spotting can be a free hobby. If you really must spend, here is a simple guide to hardware that can take your interest in this field to a higher level. More detail on each section is provided further into the book. I am a super-scrimper at heart and will always encourage spending as little as possible to achieve a goal. (I have a book partly about that too... **Get Out Of Debt Fast** – available on Amazon).

Binoculars: If you do not own binoculars, purchase a used pair. eBay is the perfect place, if you are registered, or simply ask someone who is to get a pair for you. There are thousands available at any one time. This is a perfect form of recycling. Many people have the opinion that eBay is losing jobs in the high street. This may be true, but thousands of new cottage industries are starting instead, working from home in a spare bedroom or garage etc. Delivery companies are enjoying a boom from this too. It is an unseen form of business but creates jobs with fewer overheads, and hence lower prices that benefit the consumer.

The most common problem with used binoculars is the misalignment of the internal prisms due to a hard drop on the ground. No matter how much you mess with the settings, the two images never line up and you will end up with a real headache looking through them. Before purchase, ensure that:

The binoculars are correctly aligned.
The lens surfaces are in good condition.

Ask these two questions first to cover yourself in case of fault when they arrive. Lens coatings are not too crucial for night observing as they are mainly required to reduce daylight glare.

Binocular power is expressed in two ways. For example: 8x30 means the magnification is 8x normal vision, and the diameter of each lens at the front is 30mm. The larger the second figure, the fainter satellites you can observe. Regarding magnification; the higher the power, the smaller your field of view, which makes it more difficult to find and track moving objects. In the case of looking for satellites, a lower power is best. This will give a wider field, a brighter image, and less shake as you track a satellite.

Relative Brightness: As with any optical system, there is a trade-off between image brightness and magnification. The higher the magnification, the fainter the image becomes. When purchasing a telescope, one can change the magnification with a set of differing eyepieces. However, binoculars tend to have a fixed optical system so that the relative brightness cannot change. Therefore, it is much more important to pay extra attention to this factor when purchasing binoculars.

Any astronomical instrument will give an image of a set brightness according to two factors – lens diameter against magnification. A bright image is required to detect faint objects.

Relative Brightness = Objective Diameter Squared divided by Magnification Squared.

For 7x50mm binoculars:
50x50=2500   7x7=49
Therefore, Relative Brightness is 2500 ÷ 49 = 51.

<u>**Examples**</u>

Binocular Size	R.B. Factor
7 x 35mm	25
7 x 50mm	51
10 x 50mm	25
20 x 60mm	9
15 x 70mm	22
15 x 80mm	28
25 x 125mm	25

A factor of 20 is usually the lowest figure for general astronomical use; the higher the better. An ideal pair of binoculars for a beginner is 7x50. These will give a very wide field of view and have a brightness factor of 51. Any satellite that is larger than a small wheelie bin / trashcan may be seen up to around 2,000 km up (1,300 miles). For more a more experienced observer, perhaps 15x80 or larger would be better. From the previous table, the 20x60mm binoculars are poor for astronomy.

My largest pair owned was 25x125. This allowed me to see satellites as far as around 36,000 km (22,000 miles). The binoculars I now mostly use are 12x70mm; they have a relative brightness of 34.

High-altitude satellites move very slowly across the sky, so there is plenty of time to home in and move the instrument along the track. These were not practical though for the fast moving / low orbit satellites. I did manage to catch a glimpse of the Space Shuttle Columbia on STS 9 mission of November 1983 and did see a definite shape. John Young was the commander on board – the only astronaut to have walked on the Moon and fly the shuttle. I did wave to him but I guess he was too busy to notice.

25 x 125 Binoculars by Vixen; powerful enough to see satellites up to 36,000 km. I named them Skylab after my first photographed satellite.

The figures do need to be studied to see if they are suitable for astronomy or satellite tracking in particular. 2.1° is normally too small for satellites, but I purchased these to view geo-stationary targets 36,000 km above the equator.

Telescopes: Normally telescopes are useless for most amateur satellite tracking, as the satellites move too fast to follow with a telescope across the sky. However, there are specialised areas where telescopes become essential.

At the beginning of the space age, Russia, Britain and the US produced highly specialised cameras built into telescopes to track the orbits of satellites accurately, and even attempted to image their shape. (Some of these details are still classified).

Technology of this calibre is now in the hands of serious amateurs. Satellites can be tracked accurately via software controlling the drive motors of the telescope, and then sometimes be magnified highly to reveal the shape of the satellite itself. A recording can be made via webcam, digital SLR camera or a specialised CCD camera specially built for low light level astronomical images. This data can then be enhanced and cleaned up for a final picture or video.

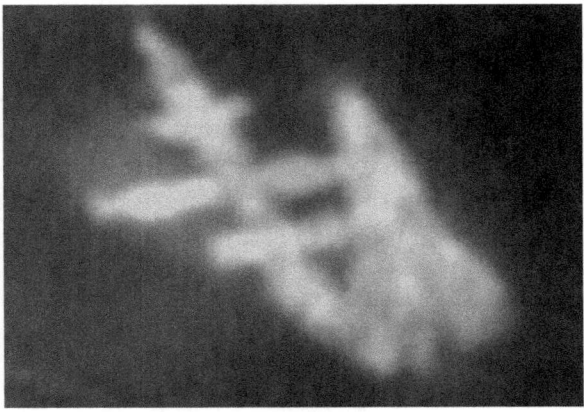

The space shuttle Atlantis docked with the Russian MIR space station in 1995. Photo *taken by an amateur satellite spotter.*

It is mentioned in the introduction of this book that satellite spotting is a safe hobby. However, if you image top-secret spy satellites and then publish these images, you could be inviting the CIA onto your doorstep within days. Even publicising the orbit data used to be enough to get you into serious trouble, although in recent years they have largely given up policing it.

This particular field of telescopic satellite tracking can be a book in itself, and will be dealt with at an introductory level later in this guidebook.

Cameras / Requirements: When film cameras were used for photography, all you needed was any Single Lens Reflex (SLR) camera, tripod and a cable release. The ideal film was a black and white HP5 or FP4. The cheapest camera for the job back then could have been purchased for around £30, namely the Zenit E. These are still available on eBay now for around £5. The big disadvantage at the time was the limitation of faintness recording. For a moving satellite, it had to be bright to be recorded on film at all. It was the same with meteors. Only the very brightest ever came out well in photographs. In addition, the film had then to be taken to a developing company (Boots in the UK as an example), whilst you waited a week before paying for that service on top of purchasing the film, even if your pictures did not come out well.

With a modern digital SLR, the whole method has changed, with massive advantages. The exact details of how to record satellites will be dealt with later, but now that the recording of fainter satellites is possible with these modern cameras, the results are instant. A third advantage is through image processing on a computer. This can improve your picture to bring out contrast and colour for a more interesting result, even with fainter objects.

From this point, we will forget film and concentrate on digital requirements. The minimum requirements of a satellite-recording camera should be:

Digital Single Lens Reflex (DSLR).

Include a standard lens (or wide angle; a bigger advantage)

A zoom or telephoto lens for more distant satellites (not essential but handy for a particular class of imaging).

Have ISO capability to at least 1,600 (ISO determines how sensitive the Imaging Chip is; the higher the number the fainter image it can record in a shorter time. Some now go up to 25,000). The Sony Alpha 7S can go up to 409,600 ISO.

At least 8MP resolution. The finer the resolution, the easier it is to record faint satellites.

Include auto exposure times of up to 30 seconds. (The shutter will remain open exposing to the sky for 30 seconds without you having to touch the camera).

Ability to lengthen the exposure time further still via Bulb setting. (B-setting: in other words, you set the exposure time manually, several minutes if you want, just keep your finger on the remote trigger button, or use a remote control button).

Have a standard tripod socket on the bottom of the camera.

Capable of either a remote control start of an exposure or cable release to avoid jerking the camera. You can get away with putting a black cover over the lens, starting the exposure by pushing the button and then pulling the black cover away gently.

This will mean that any camera shake only happens with cover on and will not spoil the result – this is technically known in the trade as the 'black hat trick'.

It is possible to obtain such a camera for around £100-£250 ($150-$375) if second hand: £250-£500 ($300-$600) new as a rough guide. I have used a Canon 600D. As of 2017, I upgraded to a Canon 1300D – with much more rapid processing time. The 600D took around 20 seconds, the 1300D – 2 seconds.

Note: SONY ALPHA CAMERAS: *If money is no object, consider the Sony Alpha range of digital SLR cameras. These have no shutter mirror that flicks up and down creating a slight vibration. The maximum ISO value can be as high as 409,600. The standard ISO in the same price range is around 24,000. You will need very dark skies though for the full facility. The built-in image processing software is excellent at reducing background noise on images set at such high sensitive levels*

Webcams: These can be attached to a telescope so that a live feed can be recorded onto a video deck, DVD recorder or Laptop computer. Later, single frames of extra fine quality can be selected to 'stack' within image processing software to produce one very fine detailed image. Virtually any quality webcam will perform fine. Some are produced specifically for astronomical recording. The Philips ToUcam Pro for around £60 ($90) and Logitech Quickcam Pro for around £50 ($75) are examples of astronomical dedicated models. They have no moving parts, so a used webcam will work very well at a much lower cost, and £20 ($30) should cover it. It's the telescope that produces the actual image to record. Such forms of hardware can last for many years.

An all-sky webcam can do what it says: record the entire sky via a fisheye lens. These are specialised and currently cost around £250 ($325). Bright satellites are within reach of such a system, which is normally used for recording meteor showers, but is just as good for satellites too. A dark sky is required for this instrument.

16

All of these items are normally listed on eBay. Otherwise, simply mention your exact requirement details if purchasing from a store. A used camera will be fine, providing every part of it is in full working order. I do try to encourage money-saving habits if you don't do so already.

Photoelectric Observations: Some amateur astronomers use photoelectric meters on telescopes to record the brightness of an image in real time. A read-out is given on a meter and is recorded second-by-second on a computer. This can be automatically converted to a graph. Uses include eclipses of the Moons of other planets. A Moon of Jupiter, for example, will apparently vanish in its orbit as it disappears behind the planet, causing the brightness of the whole image to drop so that an accurate timing of the event can be made.

For satellite observations, a dedicated telescope mount, drive system and software is needed to follow the satellite across the sky. If the satellite is tumbling in space, then the changing brightness can be recorded and an accurate rotation speed / period can be observed. Not for a complete beginner.

Radio Receiver: Many satellites transmit within radio frequencies. Amateur radio hams used to track them in the early space age days. They transmitted data relatively slowly, and a serious amateur could track a satellite, record the signal and decode it. Operational satellites today transmit at rates thousands of times faster, so this has now become a highly specialised field. It is still possible to receive data directly from a satellite, such as MeteoSat, and produce live images of the Earth. An organisation called 'Group for Earth Observation' specialises in this field – if you wish to investigate this area further.

www.geo-web.org.uk

A clip is included on the website.

Chapter 4   Comfort Observing

If time is going to be spent outside watching these amazing machines pass over your hometown, then you need to make yourself comfortable. Coldness can soon set in to your body (especially in the UK), making it both difficult to read your satellite listing and preventing you from having an enjoyable experience. You will be wishing you were inside watching a good movie instead, and your new interest could be history.

You may just wish to watch a single satellite pass just so you can amaze your friends and proudly inform them that you have seen Cosmos 2103 with its rocket case trailing and tumbling behind. In this case, just get your shoes / slippers and coat on five minutes before the predicted pass, and pop outside. Get dark-adapted, (a few minutes outside will allow your eyes to adjust to the dark), and then if you see stars you should see the satellite as predicted. This requires very little planning; you can pause your movie in the meantime and see the space age event take place in the real world rather than seeing a report on the internet or reading a 30-word article in the newspaper the following day.

If, however, you want to undertake a satellite marathon and witness dozens pass over on a beautifully clear moonless night, then a little planning will make it a much more enjoyable and memorable experience.

Clothes: In the UK or similar climate, wear a thick coat with either a hood or a separate hat. Most of your body heat is lost through your head. Thick thermal socks and boots keep your feet warm. Do not stand on concrete if possible. If there is no other option, try standing on an old piece of wood. Wear thermal gloves unless you are setting up your camera, which can be a delicate operation. Do not forget glasses if you need them for adjusting your camera and reading from your satellite list.

Deck Chair: If you are looking for faint satellites in high orbit that require binoculars, do consider a deck chair. This will allow long duration observing sessions in comfort, and a steady position will allow accurate sweeping and observing for satellites.

Warm coat, hood, gloves, thick boots or shoes, and off the ground if possible (plus the most essential items: warm blanket, hot chocolate and biscuits - Ed). This combination produces relaxing and warm conditions for satellite spotting marathons. Position yourself in a sheltered area away from winds. The binoculars shown are 12x70.

Image of the author: ask for a signed photo of this if you desire. At least it shows my best side.

Most of the time I chose to stand, especially if I am only out to see one or two targets or if I see clouds approaching fast. The deck chair is not worth the effort.

Print-Out & Red Torch (flashlight): Use www.heavens-above.com for the satellite predictions. A dedicated chapter in this book describes how to use it. Print out the satellite list for that night and highlight the objects you most wish to see. If you want to be flash, try out the 'app' that will give live information on each potential satellite that can be viewed from your location.

Ensure you have a red filtered torch to illuminate the list. A white light will ruin your dark adaption for a while but can be preferred for reading small print. An LED torch will last and last without needing new batteries. I use solar charged batteries so the running costs are zero for years (yes I am that frugal).

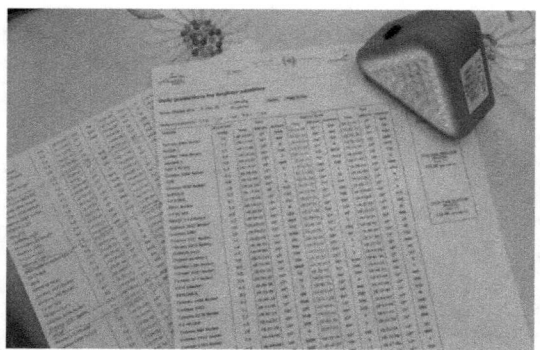

Tick off the satellites after observing them as a record of your achievement. At certain times of the year, over a hundred satellites can be seen in just four hours or so. My personal single evening record is 119 recognised satellites including seeing seven passing over at once.

This is our observing site in Arizona, USA... waiting patiently for the night's observing run. The fence is to keep out snakes.

Chapter 5   Why can we see Satellites?

We can see stars at night as they all shine by their own light. The Moon, planets and satellites, however, do not generate light but instead reflect light from the Sun. During night time on Earth, we are standing within the Earth's own shadow, and the Sun is below the horizon. Stand on a ladder and you won't expect to see the Sun. Walk to the top of a hill – still no Sun. Fly up in a plane to 4000 metres – nope! However, fly up in a rocket to 300 km or so, and aha! We see the Sun above the horizon again. It is now illuminating you, but the ground down below is in darkness with city streetlights, fires, and flares from oilrigs.

At sunset, there is a much better chance of experiencing this on a small scale if you feel athletic enough. At the back of my old school, (Upbury Manor, Gillingham, Kent, UK), there is a 40° inclined hill about 60 metres high that leads down to the town of Chatham. In November and February, the Sun sets around 4pm. I used to walk to the bottom of that hill after school on a very clear day and watch the sunset. I would then run half way up as fast as I could, turn and watch the sunset again; then finally up to the top to see a third sunset! The bottom of the hill was getting dark while the top was still lit by the Sun.

Now imagine a ladder; climb up it quick and witness a fourth sunset; then onto a small plane up at 1km; then a jet at 15km, and finally a satellite at 400km! Anybody can witness the sunset many times on the same day if you keep going higher at a faster rate than the setting Sun. The Sky TV satellite is only in darkness for 45 minutes or so per 24-hour orbit. Placed far enough away from the Earth, and it would be visible 24 hours a day while down below that part of the Earth may be in darkness. Deep space probes experience the Sun constantly.

Above: As the Sun has just set to the West, the Earth's own shadow can be observed to the East. The low dark line in the top image is the beginning of the shadow within the atmosphere. As the Sun gets lower, this line rises rapidly until we are deep inside it. The Full Moon is shown here too. The opposite effect takes place early morning at sunrise. This effect is sometimes known as 'The Belt of Venus'. The exact opposite point to the sunset is the 'Anti-solar point'.

The angle between the Sun and a satellite is called the Beta angle.

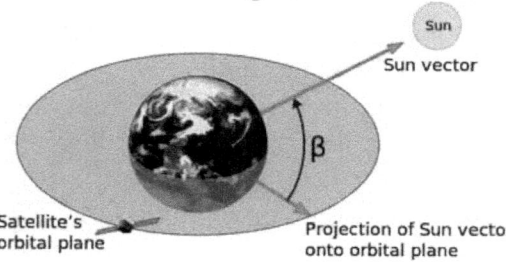

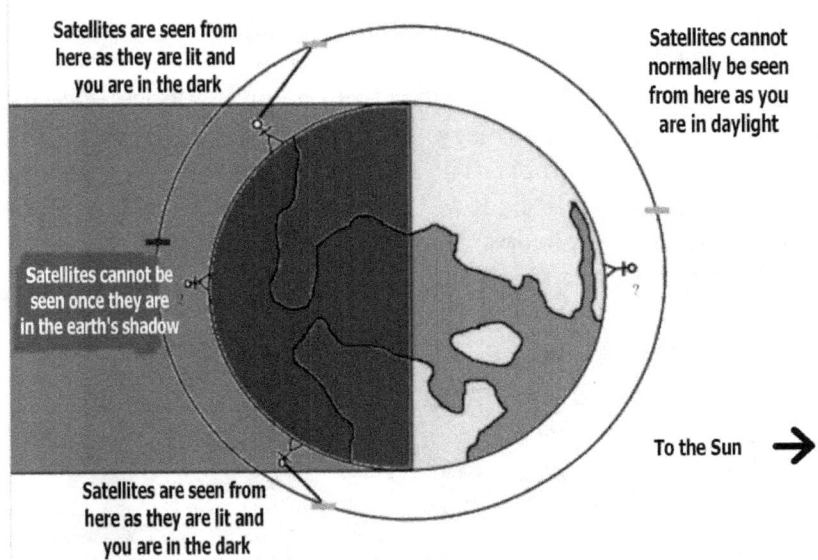

Satellites are seen from here as they are lit and you are in the dark

Satellites cannot normally be seen from here as you are in daylight

Satellites cannot be seen once they are in the earth's shadow

To the Sun →

Satellites are seen from here as they are lit and you are in the dark

Above: A diagram showing what part of a satellite's orbit would be visible from the ground.

Conspiracies: I wish to use this spare half of the page as a message to all those who think all satellites are fake. Please get out more – literally. Some people explore websites such as YouTube and absorb the conspiracy nonsense. They all have one habit in common: sitting at the keyboard and deciding what to believe by listening to others who know no better. The conspiracy rubbish is just that. They need to go to an Astronomy club and learn real science or buy this book – please recommend the book first.

If you want to have a laugh, just search *'Fake Satellites'* on www.youtube.com and see the nonsense unfold. It increases my blood pressure too much, so I refrain from it now. It is Anti-Education in its ultimate form. Some people just like to stand out in the crowd and believe in something different, but they merely result in embarrassing themselves instead. We have two books related to this: explore www.outerspacebooks.com

More on the Earth's Shadow: As the Sun is 109 times wider than the Earth, our shadow is therefore tapered as a cone. It reaches out to a point 1.36 million km from the Earth. Beyond that, the Sun is fully visible apart from a small silhouette of the Earth against the Sun's image. The shadow is in two parts: The Umbra and the Penumbra. The Penumbra is just partial shadow; only the Umbra is full shadow. Both parts are shown as the Moon passes through the shadow sometimes when it is full. This how is a lunar eclipse works.

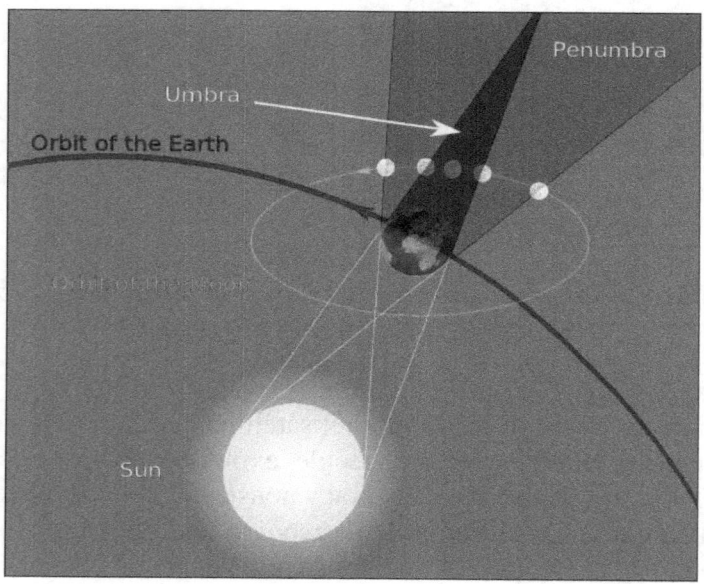

The next diagram shows the view of the Earth from space within the Earth's own shadow, with the Sun aligned in the background. At point A, the Sun is totally eclipsed. Point B will show the outer solar atmosphere, the Corona, but not the Sun's surface itself, just as in a total solar eclipse from Earth. Point C and D are at or beyond 1.36 million km from Earth. Much of the solar disc is now visible, and the Earth will just be a silhouette against the Sun's face. The same can occur on the Earth when Mercury and Venus pass in front of the solar disc. This is known as a Transit. During such rare occasions, the Earth is actually passing through the Penumbra part of those planets' shadows.

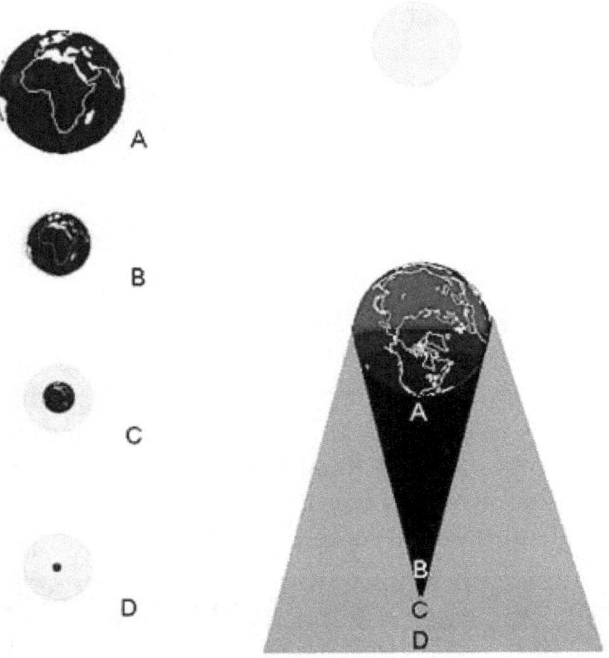

Below: The Umbra of the Earth's shadow is the deep darkening region shown here on the Moon; the surrounding hazy area is part of the Penumbra. Lunar eclipse of March 2007 from Sittingbourne, Kent, England

As a satellite travels into the Earth's shadow, it fades for a few seconds as it enters the Penumbra area, then vanishes completely as it passes into the Umbra. The time this takes depends upon the angle of approach. The satellite may head straight to the shadow at 90 ° and can fade in just 5 seconds. If it approaches from a shallower angle, it could take 30 seconds or more. On rare occasions, it may skim the edge of the Umbra and just make a fainter crossing of the sky without actually fading at all.

A total lunar eclipse will show the Moon glowing in red light. This is due to the red part of the Sun's spectrum being refracted through the Earth's atmosphere around the edge, and redirected into the Earth's shadow. The light bends slightly inward, and by the time it reaches around 300,000 km the whole of the Earth's shadow is filled with this red light. Since the Moon is 400,000 km away, it bathes in this soft light during eclipses.

September 2015 Lunar Eclipse from the UK. The Moon is almost totally immersed in the Earth's Umbra part of its shadow. The left side is showing the Penumbra.

Seasonal Changes: As the Earth orbits the Sun, its 23° tilt remains pointed in the same direction. The North axis will always point toward the North Star. However, our tilt related to the Sun will change slowly day by It's this effect that causes seasons and also alters the angle of our own shadow in the sky.

Previous: As the Earth rotates, the Moon's orbit is tilted in comparison with the Sun. The Earth's tilt alters slightly each day compared to the Sun too, due to seasonal shift. All these changes alter the Moon-rise / set position daily as well as its orientation.

Some people on YouTube think it is the Earth tipping over or the Moon's orbit changing, and that there is a big conspiracy covering it up. Nonsense! The Earth's tilt is perfectly normal, and perhaps these folks should go to an astronomy club and learn some real science rather than guessing at a computer keyboard and confusing non-scientists who watch their clips.

During local Winter, the Earth's shadow is high above us within two hours or so after sunset and two hours before sunrise. The Summer effect will keep the shadow very low, even in the middle of the night. Therefore, satellites would be seen all night with complete crossings. The next diagram is of a satellite passing over the UK in June at a few minutes to midnight.

The further you are placed from the Equator, the more this is pronounced. Close to the Equator, the shadow is above you within 100 minutes or so after sunset regardless of the time of year. It's this angle of the shadow compared to your latitude which defines the number of satellites that can be seen each night. No matter where you are, or the season, many satellites will be observed, but the number of them and whether you see them enter the shadow or not are defined by those two points: your Latitude and Date.

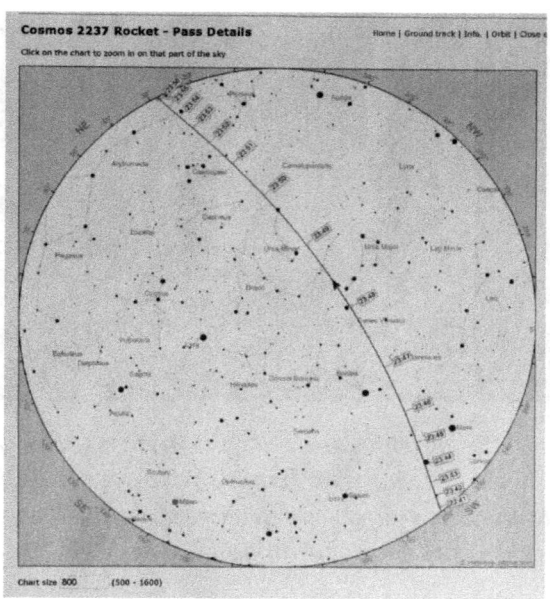

For Northern Hemisphere observers at latitudes above 30° N, the best months for satellites are from March to late September. For Southern Hemisphere latitudes below 30° S, September to March. Between 30° North and South around the Equatorial region, the shadow effect is roughly constant all year.

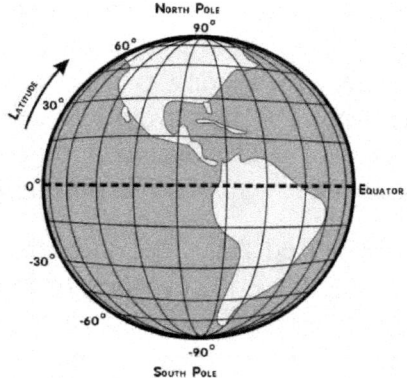

Brightness: With round shaped objects, such as stars and planets, the further away they are the fainter they appear, i.e. the greater the distance, the more the light is being scattered. An early satellite, Echo 1, was in fact a balloon that reflected radio

28

signals, and it was found that its brightness, as well as that of other roundish satellites, does in fact diminish with distance in a predictable manner, as does that of planets.

The mathematical law, the Inverse Square Law, is a very simple one to grasp. For example, if a planet is 50,000,000 miles from us then is observed again at 100,000,000 miles, the comparative brightness is calculated as follows…

50 to 100 is a factor of 2, or double. If we inverse 2 x 1, we get ½. If we then square ½, we get ½ x ½ = ¼. (More about this mathematical equation further into the book).

However, most satellites generally are not round, and extensive flat areas, such as solar panels or antennae, reflect light in a beam rather than spraying it out. Hence, the Inverse Square Law in such cases does not accurately apply, meaning that a satellite in a highly elliptical orbit may be brighter when higher up if a flat surface is reflecting straight at you.

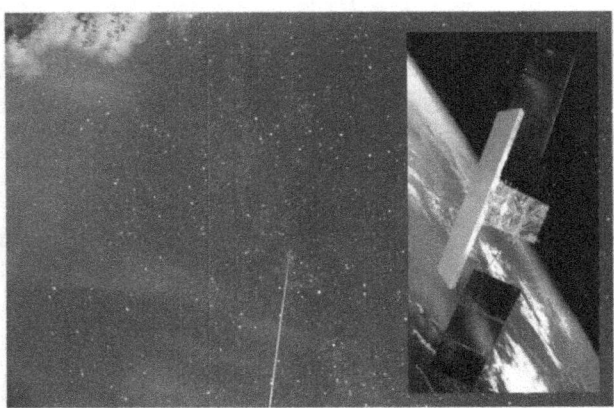

In the above image, the satellite, SkyMed 2, passes into the Earth's shadow. These fading effects are fascinating to observe. SkyMed can produce very bright long-lasting flares, as it did here. Normally, this satellite is around Mag 3.5, but here we see sunlight being reflected perfectly off the flat antenna, which increased the satellite's brightness to Mag 1.
Photo by the author in October 2014 from his own garden.

Chapter 6 Terms & Types of Orbit

Sub (Partial) Orbit: The V2 rockets in WW2 achieved the first true sub – or partial – orbit to the edge of space. At Wallops Island USA, Fairbanks Alaska, Norway, and Canada, rockets are still sent on sub-orbital hops to the upper atmosphere to test new satellite hardware and conduct research on the Northern Lights, magnetic field and atmosphere etc. Such vehicles are known as Sounding Rockets. Many amateur rocket builders have achieved sub orbits that carried cameras to the edge of space.

Low Earth Orbit (LEO): The lowest orbit possible is around 100 km. After just one trip around the Earth, the atmospheric drag will slow the satellite down enough to prevent a second complete orbit. This is the new official height limit where space begins, and where the Virgin Galactic service reaches its highest altitude, still being sub-orbital.

The lowest orbit that is practical is 160 km and is completed in 88 minutes. The maximum altitude described as LEO is 2000 km, making a 127-minute circuit. The majority of visible satellites (without the use of binoculars etc.) are within this range.

Mid Earth Orbit (MEO): Around 15% of satellites are in MEO. This is a range from 2,000 km and below 36,000 km. As they are so high up, they will be faint to observe – though not impossible. Binoculars will show many of them. At 22,200 km, an orbit takes 12 hrs, ideal for navigation and communication satellites.

The first MEO satellite was Telstar 1, launched in 1962. At 950km altitude, it was the first TV and telecommunications satellite. Mainly designed to serve between the US and Europe, the receiving station in the UK was at Goonhilly Downs, Cornwall. Although no longer functioning, the satellite can still be observed, as the orbit won't decay for at least another 1000 years! If you use the Heavens-Above website, (there is a dedicated chapter), the details of visible passes will be given from your location.

At 22-24,000 km, this position is reserved mainly for the Global Positioning Satellites, (GPS). From the ground, such satellites will travel slowly across the sky, making it easy for a SatNav (Street-pilot) to receive signals from and calculate the ground position. The new Russian Galileo GPS is being launched to 23,200 km to avoid any potential collisions with the US GPS. A group of satellites performing the same job is a Constellation. China is launching their own system. As of June 2014, they were negotiating with Russia in combining the signals between two systems for an even more reliable service.

High Earth Orbit (HEO): All Earth orbits above 36,000 km (Geo Stationery) are considered HEO. These all take more than 24hrs to circle the Earth, so instead of travelling generally West to East, these *seem* to go East to West! This is only due to the fact we are rotating in a faster time than the satellite.

LELA 1A was the first such example. In 1965, it was placed at 118,000 km - above the Earth's Van Allen Radiation Belts. The satellite was designed to observe Russia and ensure President Kennedy's 1963 Nuclear Test Ban Treaty was being honoured. If the satellite orbited within the Radiation Belts, then false data would have interfered with the observing and render the project useless. It was designed to last 6 months, but lasted 5 years. The last such satellite, VELA 9, was launched in 1969. It operated until 1984. The Cold War came to an end, and so that was the last of the series.

The VELA 5 satellite in the clean room shortly before placing in its rocket. NASA

Highly Elliptical Orbit: These orbits are generally below 1000 km at minimum altitude and above 36,000 km at maximum. Some communication satellites use this kind of orbit, as the majority of its path will appear to remain almost stationary in our sky. Some are achieved by accident if a final firing of a thruster fails and leaves the satellites in an unintended elliptical path around the Earth.

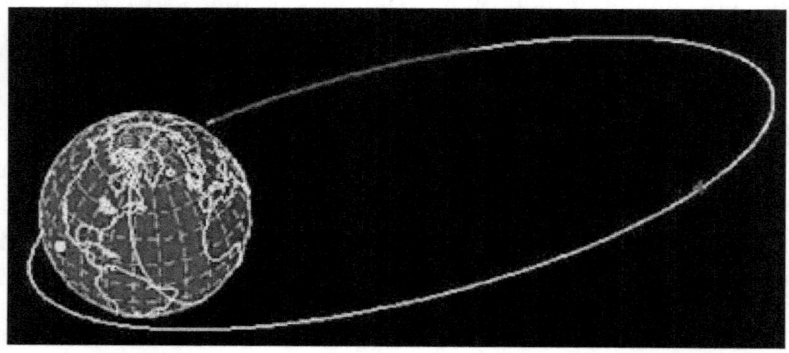

Geo – Stationary Orbit (Clarke Orbit): The higher the orbit, the longer it takes to complete a circuit of the Earth. At 36,000 km it will take 24hrs for one single orbit. As the Earth also rotates in 24hrs, the satellite, once positioned, will remain above the same point over the Earth forever.

Such satellites occupy a single ring above the Equator only. The reason for placing these satellites apart is to avoid radio-frequency interference during operations, which means that there are only a limited number of 'slots' available. Only a small number of satellites can be operated in geo-stationary orbit. This has often led to conflict between different countries wishing access to the same orbital slots, (countries near the same longitude but differing latitudes), and radio frequencies. These disputes are addressed through the International Telecommunication Union's allocation policy. In 1976, eight countries located on the Earth's equator claimed sovereignty over the geo-stationary orbits above their territory. However, these claims gained no international recognition.

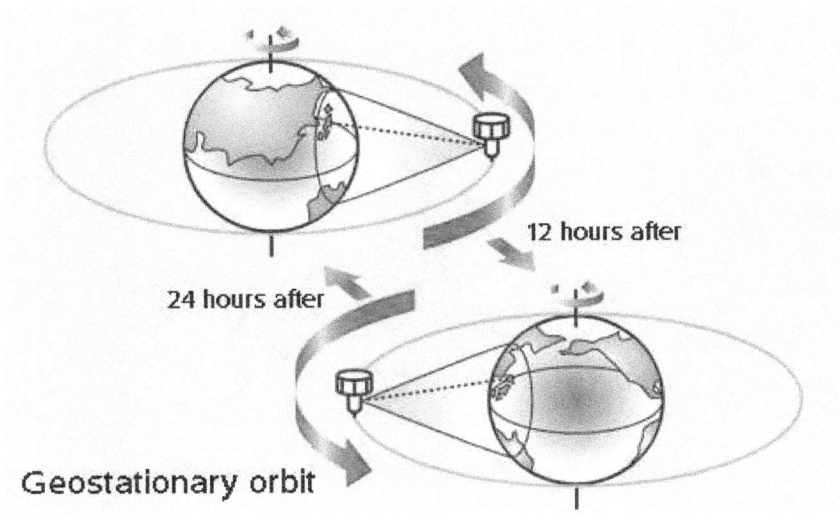

12 hours after

24 hours after

Geostationary orbit

All satellite dishes like these point to a geo-stationary satellite such as Sky above the Equator. Dishes were much larger up to the 1980s, as radio wavelengths were mostly used. However, these frequencies are contaminated by background noise from stars etc., so a stronger signal was required. If you observe the sky in Microwave instead, it is almost completely dark. Most satellites from the 1980s used Microwave transmitters, and so smaller and more practical dishes could be employed.

Microwave receivers are smaller than radio as the signal suffers from less background noise. Lower power transmitters and receivers become possible.

Satellite radio receivers in South Dakota.

The angle of the dish will depend upon your Latitude. The closer you are to the Equator, the higher the tilt. This alone proves the world is a globe. If you are fan of YouTube, try searching *Flat Earth* clips. YouTube is a great site, but does attract all kinds of nice people with strange and completely wrong ideas. These are doing a terrible disservice with regard to educating youngsters, and conflicts with what they are being taught correctly at school. This amounts to 'anti-education'. Incidentally, flat-earthers can be found all over the globe ...

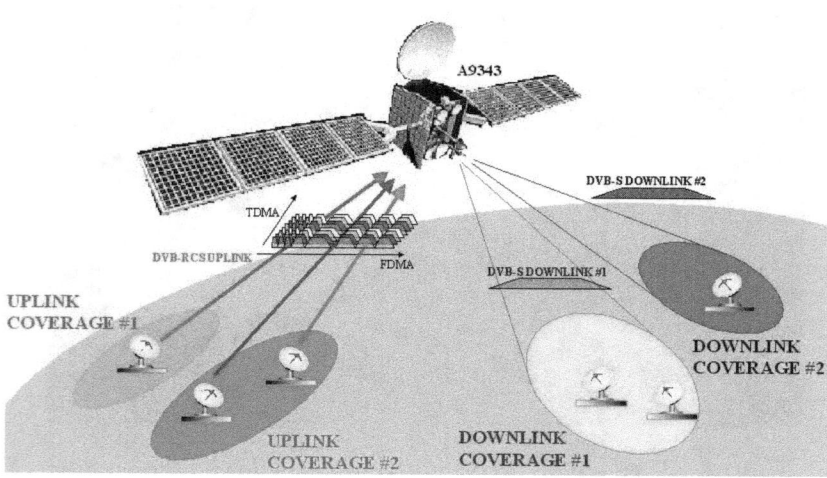

For a single service to cover the whole world, three satellites 120° apart are required.

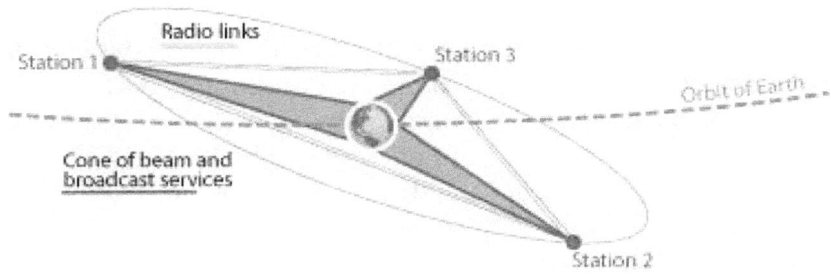

When they run out of thruster fuel, the satellites are at the end of their useful life, as they are no longer able to remain in their allocated position. The transponders and other on-board systems generally outlive the thruster fuel, and by stopping N-S station keeping, some satellites can continue to be used in inclined orbits where the orbital track appears to follow a figure-of-eight loop, (centred on the Equator), or else be elevated to a 'graveyard' disposal orbit further out.

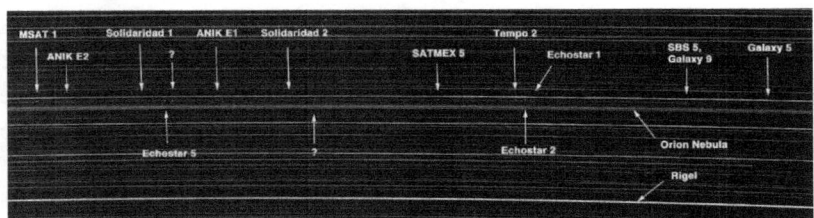

Geo-stationary orbiting satellites; the camera is kept still while the stars trail as the Earth rotates. Picture by Bill Livingstone of Tucson, Arizona.

Equatorial Orbit: An orbit that passes constantly above the Equator only. All Geo-stationary satellites are in Equatorial orbits. These are best achieved from a mobile launch platform (Ocean Odyssey) placed directly on the Equator to save on fuel and complex orbital manoeuvring.

Inclined Orbit: The inclination of a satellite's orbit is the angle that the orbit crosses the Equator. If a satellite has a 0° inclination, then it would be orbiting directly over the Equator. If a satellite has a 90° inclination, then its orbit is at right angles to the Equator and it would pass over the poles instead. All inclined orbits are therefore between 0 °and 90° to the Equator.

Polar Orbit: The satellite passes over the North and South Poles on each orbit and eventually passes over all points on Earth as the planet rotates beneath. A pass can be seen as North to South, but after a few orbits the same satellite will be seen as South to North, then back again, apparently reversing direction. However, this is only due to the fact that the Earth has rotated, so what the observer will see is merely the opposite side of the orbit, not an actual reversal in direction.

Sun-synchronous Polar Orbit: This is a special kind of polar orbit. When travelling within this orbit, a satellite not only travels over the North and South Poles, but it passes over the same part of Earth at roughly the same time each day, which is essential for some of the daylight observations.

Polar Orbit: The satellite passes over the North and South Poles on each orbit and eventually passes over all points on Earth as the Earth rotates underneath in this orbit. A pass can be seen as North to South, but after a few orbits, the same satellite will be seen as South to North then back again. This is only due to the fact the Earth has rotated so the observer will see the opposite side of the orbit and not that the orbit has reversed direction.

Sun-synchronous polar orbit: This is a special kind of polar orbit. When travelling in this orbit, a satellite not only travels over the North and South Poles, but it passes over the same part of Earth at roughly the same time each day; essential for some daylight observations.

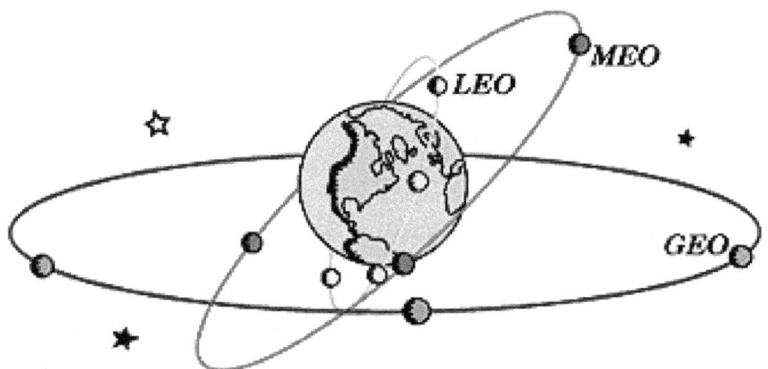

Exploration of the Solar System is performed by what are called 'probes' rather than 'satellites' – unless of course they get into orbit around another planet. Relevant clips are on the website. Just look up the page, **'Types of Earth Orbit'**.

Other Orbital Terms: A few extra terms need a little background on to ensure a fuller understanding of satellite orbits. This will allow any reader to access other sources and follow them with ease and clarity.

Apogee: The point in the orbit of the Moon or of an artificial satellite most distant from the centre of the Earth. Also the point in an orbit most distant from any body being orbited. It is also the slowest part of an orbit. For easy remembrance, 'A' is for 'Away', meaning that 'Perigee' is therefore the nearest point.

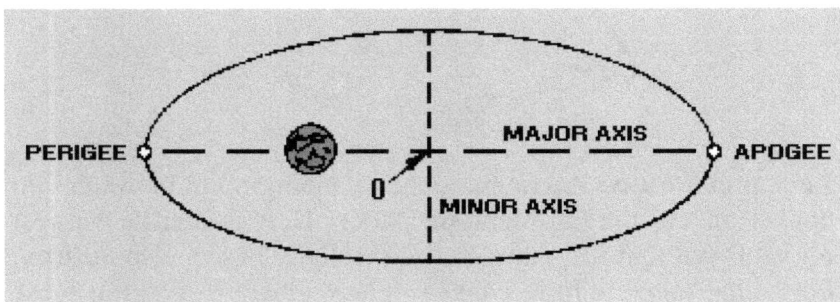

Delta V: Simply the measurement of a change in the velocity of the spacecraft – not velocity itself, and is caused by an engine burn or the after-gravitational influence of another body.

37

Eccentricity: Every orbit is an Ellipse, not a circle. A circle has one centre; an ellipse has two. The radius of an ellipse has two figures, the major axis and the minor axis. The proportional difference between them is Eccentricity.

Equatorial Plane: An imaginary line above the Earth's equator on the sky. Geo-stationary satellites are positioned there.

Equinox: This occurs twice a year when the Earth is exactly sideways on to the Sun. We get 12hr of day and 12hrs of night across the globe. The shadow is perfectly in line above the Earth's equator. 21st March and 21st September is usually when this happens, depending upon the timing to the Leap Year.

Inclination: The angle between the Earth's equator and the orbit. Expressed as *'i'*, this angle, together with altitude, will determine whether an observer will witness the satellite at a given Latitude or not.

Lagrange Points: A gravitational balancing point between two bodies such as the Earth and the Moon. Here a satellite can rest for years without firing thrusters too often. These are so far from Earth that there is little chance of any amateur imaging these satellites.

Latitude: The number of degrees north or south of the Equator. The Equator is 0°N or S and the North and South Poles are 90°N and 90°S.

Longitude: The number of degrees east or west of the Prime Meridian, (the 0°E or W line going through Greenwich, England and the North and South Poles). The Prime Meridian divides the globe into eastern and western hemispheres. It runs between the Poles through the Pacific Ocean on the side of the globe opposite England. This line is the International Date Line. Degrees longitude are 0°E or W at the Prime Meridian and 180°E or W at the International Date Line.

Mass: The weight of any stationary object on Earth in kilograms. Take it into orbit, and the weight is now zero, but the Mass will remain the same, (because if you brought it back down the original weight resumes).

Orbital Decay: The reduction in altitude of a satellite's orbit, caused by drag from the thin atmosphere or solar wind. Gravity does the rest.

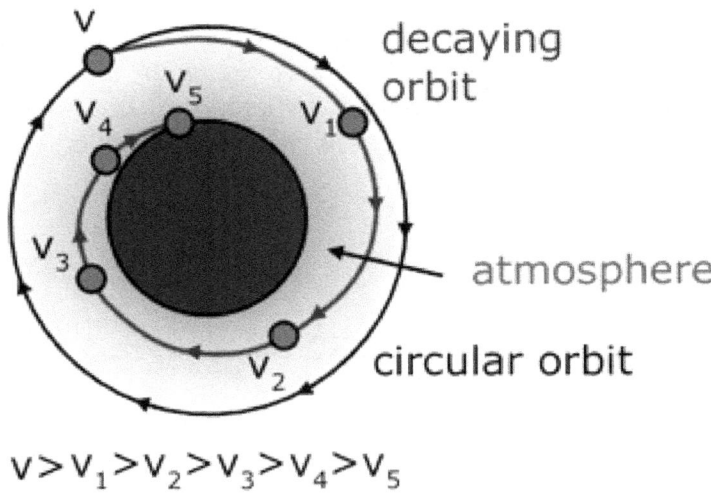

$$v > v_1 > v_2 > v_3 > v_4 > v_5$$

Orbital Period: The time it takes a satellite to complete one orbit.

Parking Orbit: The orbit that is intended for operations or for temporary use until ready for a transfer orbit to a higher altitude or for leaving the Earth.

Perigee: The point nearest the Earth's centre in the orbit of the Moon or a satellite. This is also the fastest part of an orbit.

Pitch Axis: Satellite axis that is at right-angle roll-yaw plane.

Roll Axis: Satellite axis that points along the direction of travel.

Station Keeping: Once an orbit is achieved, the satellite commences its job description. During every orbit, it will encounter the slight tug of the Moon and the Sun; the higher the orbit from Earth, the stronger the effect. As time passes, the satellite can drift either ahead or behind its useful path, or even north or south in its designated inclination. Small thrusters will fire every so often to correct this continuing error. This will limit the lifespan of the satellite as it can only carry a finite amount of fuel. Experimental systems that involve an Ion Drive can lengthen the lifespan as they use so little fuel.

Sun-out, geo-stationary orbiting satellites occasionally pass in front or very close to the Sun in the sky. The Sun is very noisy in the radio spectrum and often interferes with the satellite transmissions. This is 'Sun-out.' Microwave transmissions hardly suffer from this effect, as the Sun is quiet in the microwave frequencies.

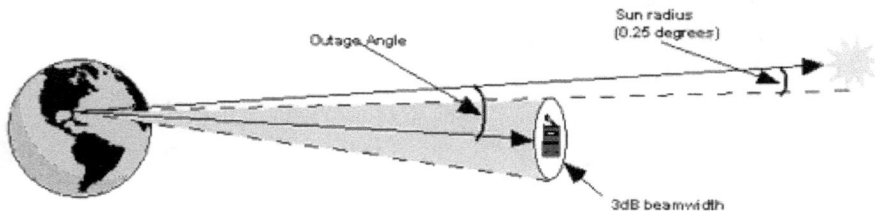

Surface Circular Satellite is when the orbital height is equal to the earth's radius – 6000 km.

Hohmann Transfer Orbit; a path from one orbit to another. This can be made by a single burn or over several burns of a gradual increasing altitude.

Yaw Axis; A satellite axis that points to the Earth's centre.

Chapter 7   Ground Track

As the Earth rotates, the satellite is revolving around too. If it is not in an Equatorial Orbit, then gradually the satellite will pass over different parts of the globe each time it gets back to the starting point and begin another circuit. The point directly under the satellite on Earth is the *Subsatellite Point*.

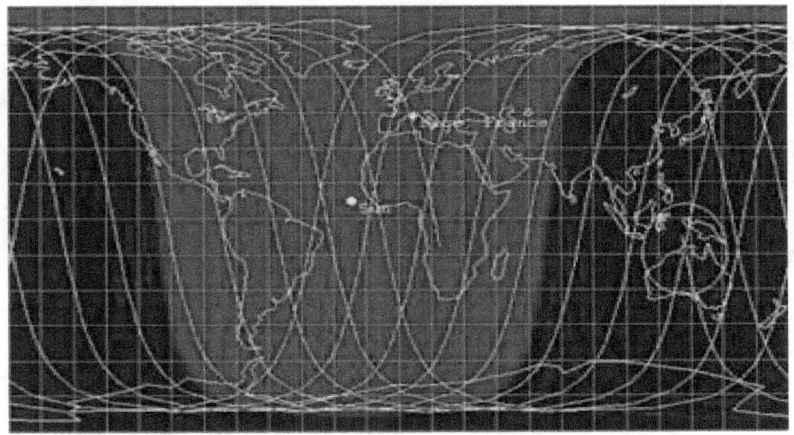

After many orbits, the ground track roughly repeats itself. Most satellites travel roughly West to East; with the Earth's rotation. This example is of a Polar Orbiting Satellite. The Earth map is usually shown as a Murcator Projection.

Many satellites can be seen twice in one evening from around April to August in the Northern Hemisphere and August to April in the Southern. The same satellite can pass over again after 100 minutes or so but as the Earth has rotated by 15° during that time, it will appear in a completely different part of the sky.

A straight line from the satellite to the Earth's centre defines the ground track; where it reaches the Earth, this is the moving ground track. The inclination of the orbit will define the distance reached from the equator. If it is at 35° from the equator, then the ground track will reach a maximum of 35° above and below the equator.

In theory, a geo-synchronous satellite will have a point on Earth as the ground track. However, perturbations will make it move around a little; usually in a figure '8' for most. This is caused by the Moon's gravity. The satellite will encounter a stronger pull as the satellite moves between the two bodies once a day.

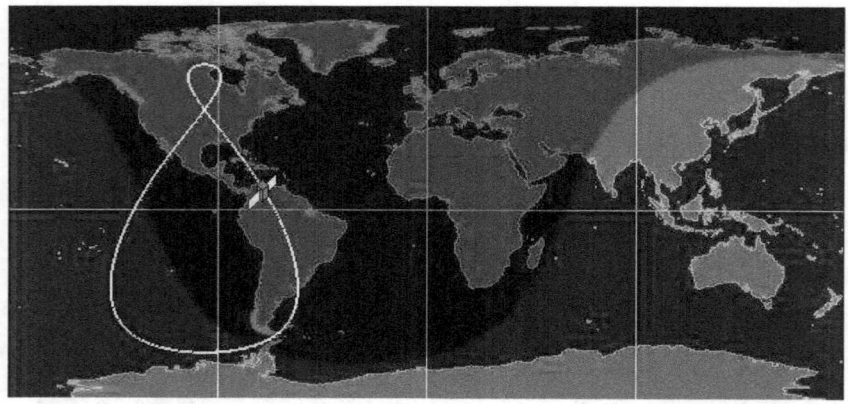

The tracks of satellites appear curved as the globe map of the Earth is translated into a flat map; Murcator Projection. It is a geometrical effect of map display rather than real.

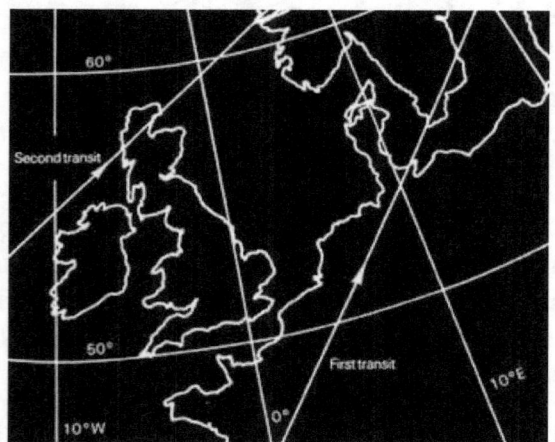

This example satellite can be seen throughout the UK twice. If the satellite was 500 km up and the observer was in London for example, the first pass would be around 60° up to the south and the second 35° up toward the north. Both will travel roughly west to east.

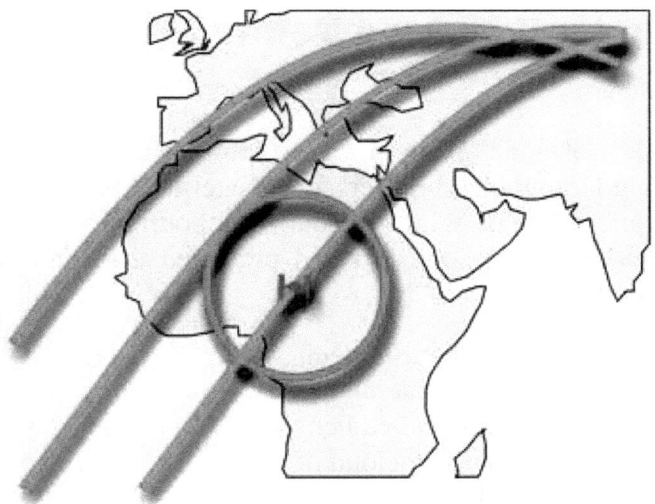

This diagram shows the ground track of the International Space Station on three typical orbits. Every year the station is pushed into a higher orbit as it slowly falls due to thin atmospheric drag. The timing of each pass then changes. The altitude can drop by as much as 50 km a year; without such station-keeping it will fall out of orbit as Skylab did in 1977.

Probably the most unusual ground track is on a highly elliptical orbit. As the satellite increases its altitude, it will physically slow down. If it is high enough, the Earth's spin can overtake the speed of the satellite and make it appear to travel backwards in the sky for the Apogee (peak) part of its orbit.

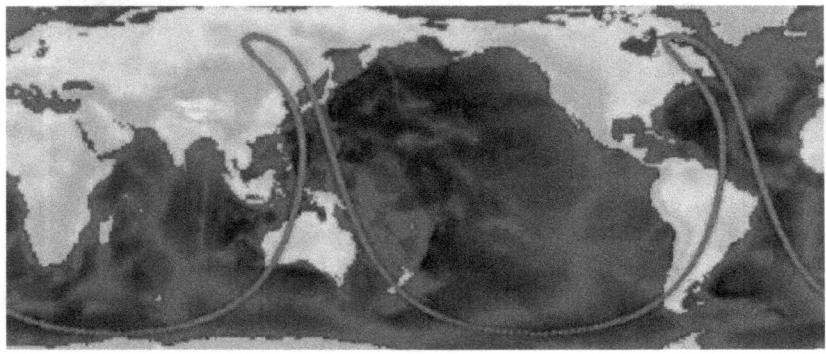

Chapter 8   Space Debris & Numbering

All kinds of junk exist in Earth orbit. Thousands of items including large rocket boosters, dead satellites, panels, tool kits, gloves and flakes of paint litter our neighbourhood. Satellites on rare occasions have collided and produced thousands of new pieces of debris.

The deepest worrying case of all was an anti-satellite weapon system tried out by China in January 2007. It destroyed an old dysfunctional satellite, called Fenyung-1C and created a massive uncontrollable expanding cloud of debris that is now a hazard to other satellites worth billions of pounds / dollars in an already crowded orbit 800 km up. They were told off and promised not to do it again; but this problem will persist for decades. As time passes, this debris field will drop in altitude and eventually even cause a threat to the ISS; it will pass through the field twice per orbit.

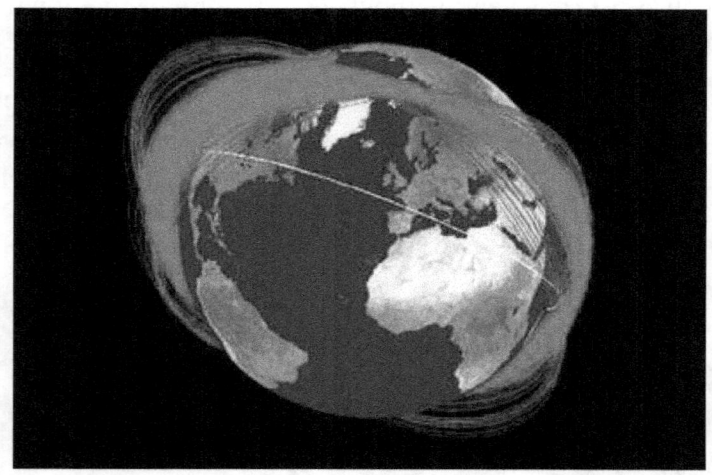

The debris cloud tracked by radar and visually by North American Air Defence Command-(NORAD) to alert active satellites and sometimes have to change orbit to reduce the chance of a collision.

Clips related to this subject; www.outerspacebooks.com

Kessler Syndrome: Donald J Kessler first proposed a depressing and dangerous scenario in detail. He worked at the Johnson Space Center near Houston, Texas within NASA's Environmental Effects Project Office. He developed what is now known as the Kessler Syndrome; the proposal that collisions become increasingly likely as the density of space junk is increases in orbit around the Earth, and a cascade effect results as each collision creates more debris that can cause further collisions and so on. Kessler first published his ideas in 1978, in an academic paper titled "Collision Frequency of Artificial Satellites: The Creation of a Debris Belt. The paper established his reputation and NASA subsequently made him the head of the newly created Orbital Debris Program Office to study the issue further. It was this that inspired the story behind the British hit movie 'Gravity.'

Early signs of this scenario predicted that there would be an increase in unknown satellite failures. Most satellites that cease functioning do so for a known technical cause by studying the telemetry data on the satellite's condition. But unexpected failures are likely to be due to debris hitting a vital part of a satellite. Europe's largest ever Earth-study satellite, Envisat, ceased functioning for no known reason on 12 April 2012. It cost over £2 billion and built in Stevenage, England. Other similar failures have followed; all in low Earth orbit.

Another predicted sign will be collisions of complete satellites; the very first confirmed one was on 10 February 2009 between Iridium 33 and Kosmos 2251 over Siberia. Both were operational at the time. As they went off line at exactly the same time, it was not rocket science to deduce what may have happened. The two orbits were studied and realised the historic dreaded truth. Over 1,700 pieces are now tracked adding to the artificial cloud that surrounds our planet.

On February 15 2015, a satellite called DMSP-F13 experienced a sudden spike in temperature followed by an unrecoverable loss of attitude control. The Joint Space Operations Center at Vandenberg Air Force Base, California, identified a large debris field near the satellite. This was almost certainly due to something hitting a vital fuel line or fuel tank.

If the Kessler Syndrome takes place in its ultimate form, then this entire region of space will become unworkable for centuries until solar radiation and the upper Earth's atmosphere naturally decays all this material on re-entry. The chain reaction will be unstoppable and alter our daily lives back to the 1960's. (Some may think of this as a good move).

The following diagram is an example of how easy it is for debris to build up in orbit. The first peak was a collision by two satellites; the next was the deliberate destruction of one by China testing their anti-satellite weapon. There is a potential chain reaction of one piece of debris causing the destruction of other satellites and so on; The Kessler Syndrome.

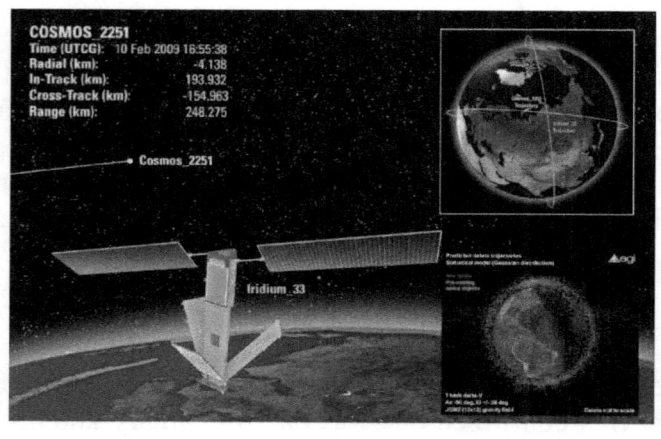

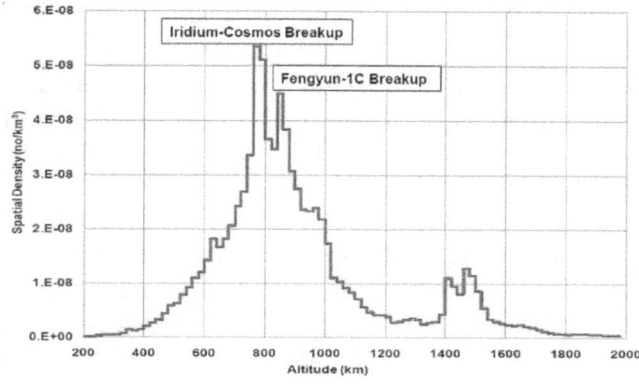

Possibly the largest risk of increasing the debris figures in orbit is through anti-satellite weapon tests (ASATS). As of 2019, four countries now have the capability of destroying a satellite up to at least 500 miles. The latest nation to join the ASAT club is India.

This debris cloud was predicted to have all entered the Earth's atmosphere within weeks. After six months, much of it is still in orbit, spreading further out and posing an increasing risk of hitting an active satellite. None of this material can ever be gathered back or controlled.

The ISS is on constant alert from NORAD regarding a potential collision. Astronauts conduct several manoeuvres a year to reduce the chance of an impact. They are also trained to respond to a call known as 'Conjunction Red.' This is where there is no time to calculate a burn to alter orbit as this can take a day or two of preparation. Instead, the crew evacuate to a Soyuz craft, power it up ready for an emergency undocking and then landing. If any object from the size of a cricket ball hit the Station, then it will almost certainly depressurise within seconds. The Space Station will 'unzip' at the weaker points and begin a chain reaction rather like a bursting balloon.

On a lighter note, the more interesting form of debris is spinning panels. The flat pieces that are tumbling end over result in a 'flashing' satellite. This next example was the casing of an Indian satellite that passed over Arizona in August 2019. As it rotated out of control, the reflecting surface changed orientation and so produced a gentle fading in and out path across the sky. The star cluster on the left is called M45 in Taurus the Bull. *(Image by the author).*

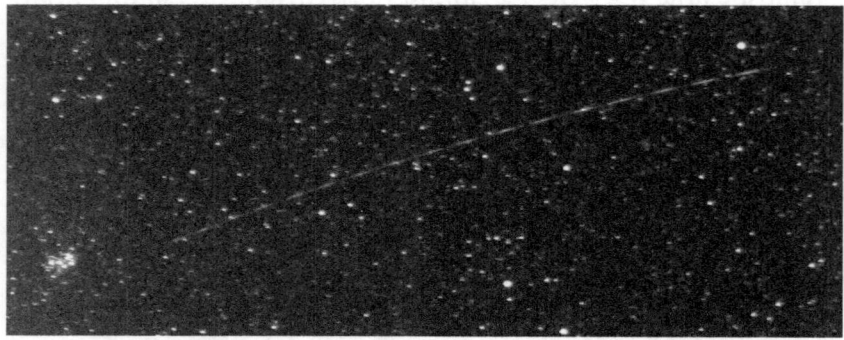

Some of the tougher form of debris will suffer from orbital decay, enter the atmosphere and land. Below is a US PAM module that landed in the Sahara and purchased by a space junk collector. It may be about time we walked around with helmets on; although it would not have been a lot use with this one...

As of 2019, 28,000 pieces of debris are being tracked by NORAD; most are larger than a cricket ball. It is estimated that over 500,000 objects, too small to track, are also potential hazards that could easily destroy an operational satellite. They all orbit at speeds of around 22,000 km per hour. An estimated 100 million pieces of paint are also in orbit that could crack a window on a spacecraft.

A flake of paint in orbit hit the main communication antenna of the Hubble Space Telescope. At an estimated collision velocity of around 15,000 km per hour, much damage can occur and can eventually render the craft useless if it hit a vital component. NASA image.

Huang Yijin and Zhang Hejin made a surprise discovery as they were walking through a forest near the village of Wuchang, Fujian province in China when they stumbled across an object that looked wildly out of place. They had no knowledge of satellite debris and concluded that it was a crashed UFO. They called the police who in turn called the China Space Agency.

The 'UFO' turned out to be the nose cone of the Chang'e 4 rocket that launched a probe to the Moon just two days earlier on 20 May 2018. The satellite was a relay communications system ready for the far side landing of a rover that took place in January 2019. After a short stay in a local grain store, the nose cone was moved briefly to a museum before being reclaimed by workers from the launch centre.

By analysing the previous cases, it is clear the Kessler Syndrome has actually begun. If no further satellites are launched from this day on, collisions will still take place, and debris will increase from say 10 million destructive items, to a billion within a decade. Every satellite will be hit in such a dramatic way that it will be rendered useless. The only solution is simply to reduce the number of objects in orbit as soon as possible. Then this scenario disastrous can be avoided.

The British company, Surrey Satellite Technologies Ltd, is amongst the world leaders in this field. A system has been invented to approach a satellite, wrap itself around and this in turn increases drag. Another is to literally harpoon a satellite.

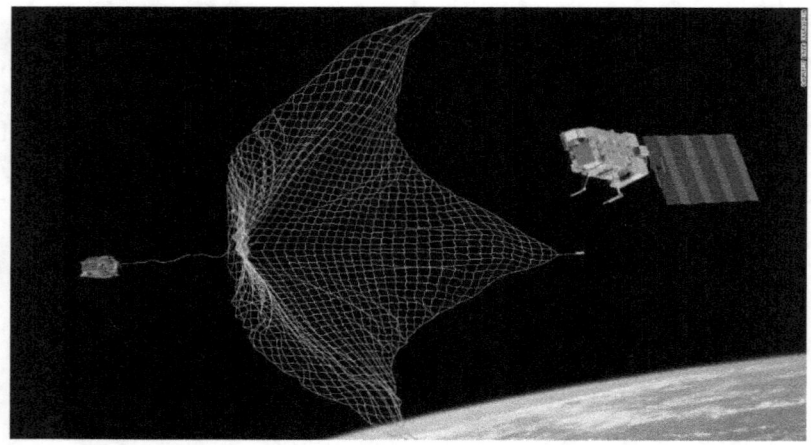

Artwork by David Duorosiesa

Numbering of Orbiting Objects: Two catalogue systems are in use; the International Identification Number (IIN) sometimes also called International Designator (IN) and the Space Catalogue.

IIN / IN : Prior to 1963, the IIN of an artificial orbiting object was a combination of Launch year followed by a Greek letter representing the number of launch that year. There are only 24 letters in the Greek alphabet, so as launches became more common it did not take long before the letters ran out. A second letter was then added and the system became cumbersome.

The IIN from 1963 is now the launch year, numbered launch by date order, plus a letter representing the part resulting from that launch. Each rocket can take several satellites and can leave large debris parts such as casings, nose cones, boosters etc. The main satellite(s) are named A, B and so on while a casing will be C, D...

If a serious failure takes place in orbit and the upper stage explodes, dozens of pieces can now enter orbit and each traceable part will be numbered AA to ZZ. Letters I & O are not used to avoid confusion as with UK license car plates.

The Hubble Space Telescope is designated IIN 199037B. 37A was the shuttle Discovery (and I saw it launched too).

Space Catalogue: As an addition with INN, another number is given to each piece of material that is larger than 10cm across. Sometimes extra items are discovered from a launch that will generate a new five-figure number. The US Strategic Command at Cheyenne Mountain Complex in Colorado defines these.

NORAD satellite monitoring centre- US Air Force image

Chapter 9   How Bright? (Magnitudes)

How bright is a star, a planet or a satellite? How can we measure or express the difference in brightness? There are two forms of expression.

1) Absolute brightness; stars are placed at an imaginary distance of 32.6 light years or exactly 10 Parsecs. Then a calculation using the Inverse Square Law is performed to determine how bright that star will be seen at that distance instead of its actual distance. A very bright star in the sky may be bright because it is relatively close; however, a faint star in the sky could actually be one of the most luminous stars in the entire universe but just seems faint because of extreme distance. This system allows astronomers to have an impression of how bright a star really is compared to others.

This form of measurement is not required for satellite spotting.

2) Apparent Brightness; this is a measurement of how bright an object <u>seems</u> to be to a human eye regardless of distance. The lower the number, the brighter the object; it may sound back-to-front but that's the way it is. A very bright object such as the Moon can be minus something, a very faint object is always plus something.

This is all we need to learn for satellite spotting. Phew!

The formula for determining the apparent brightness of any object compared to another has to be understood before you can go further...

$$\frac{f_2}{f_1} \; = \; 100^{(m_1 - m_2)/5}$$

or

$$\boxed{m_1 - m_2 \; = \; -2.5 \log_{10}\left(\frac{f_1}{f_2}\right)}$$

Just kidding, do not worry about that if you do not fancy it. All you need really to learn or at least appreciate are certain comparisons. Some of these following figures vary slightly over time as planets move around and some stars vary in brightness. Just keep referring to the next table to get the hang of it.

Object	Magnitude
The Sun	-26
The Moon	-12
Planet Venus	-4.5
Planet Jupiter	-2.5
Sirius (brightest star at night)	-1.4
Rigel in Orion	+0.1
The star Vega	+0.3
The star Aldebaran	+0.8
North Star (Polaris – 46th brightest star)	+1.4
Planet Saturn	+1.5
Dubhe in the Great Bear	+1.8
The Star Kochab in the Little Bear	+2.2
The star Pherkad in the Little Bear	+3.0
Andromeda Galaxy	+4.8
Faintest naked-eye stars	+6
Faintest stars through 10x50 binoculars	+9
Faintest stars through 25x100 binoculars	+11
Faintest ever recorded by the Hubble	+32

To photograph a particular satellite below Mag 6 (invisible to the human eye) requires perfect timing and a good camera with high ISO capability (more about that later).

This is a CubeSat; just 50cm x 60cm x 60 cm invisible to the human eye but just visible near the centre of this picture taken from Sittingbourne, Kent UK by the author.

Northern Hemisphere Observers: The next diagram is a simple map to show how to find the North Star (Polaris). Look for the 'Big Dipper.' In the UK, it is visible all year round. It is also known as the Plough, Saucepan or the Great Bear, Ursa Major. In the autumn and winter, it will be low down as in the picture below. During spring and summer, it will be virtually overhead. The shape will always remain the same, only its position will alter as the Earth spins and orbit the Sun. The stars Merak & Dubhe will always line up with the North Star or Polaris. True North is directly under it.

The four main guide stars for brightness comparisons can be;

Polaris Mag 1.4
Dubhe Mag 1.8
Kochab Mag 2.2
Pherkad at Mag 3

(Mag = Magnitude). These are ideal as the brightness are wide ranging, plus from a latitude above 35° or so, they are visible all year round. From lower down on the planet, the stars alter from one season to the next.

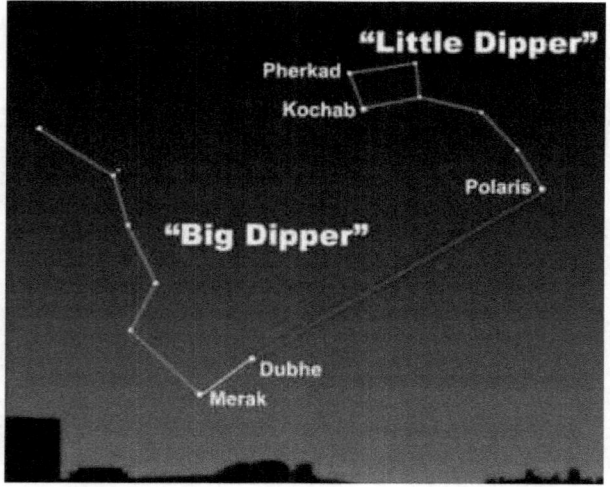

These stars can be used any night to compare the brightness of a passing satellite. If a predicted satellite is going to be Magnitude 4 (Sometimes expressed as Mag 4), you know that it is going to be a little fainter than Pherkad at Mag 3. Consequently, if another is predicted to be Mag 1, then it will be a little brighter than Polaris. A guide to a satellite's brightness will enable an observer to make a decision on whether to observe it or not.

During the night, the Earth is still rotating; as the months pass, we are also orbiting the Sun. This combined motion will change the position of the constellations but not the appearance.

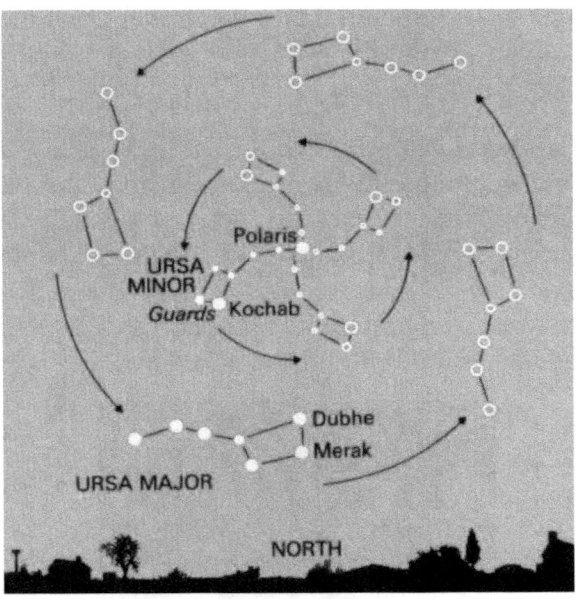

"The North Star (Polaris) is important but not very bright; like some people." President George Bush... I will not say a word.

Other conditions also change constantly. A large Moon may be in the sky and you may only just be able to make out Pherkad at Mag 3. If so and a satellite is going to be around Mag 4.5, then forget it. Go back inside and warm up. It will be too faint to see that night unless you use binoculars. The Moon orbits the Earth in a 29.5-day cycle of phases. Check a Moon calendar to see when it is visible in the morning instead of evening, and then the sky will be much darker to allow Mag 4.5 satellites to be seen. If you aim to use binoculars, then a Mag 4.5 satellite will be seen but not with the naked eye. All these examples are assuming you are just looking with your eyes.

Equatorial Observers: If you live within 3000km north or south of the equator, other guide stars will be required as a guide to brightness. Further South in New Zealand for example then there are stars that are visible all year as with the Great Bear in the UK.

No stars are visible on the equator all year. As the Earth spins and we orbit the Sun, the stars constantly change. The Earth's shadow also rises rapidly so the window for seeing them after Sunset can be shorter than at latitudes further from the equator. The added advantage though is that every single satellite in orbit can potentially be seen as they all cross the equator at some point.

Southern Hemisphere Observers: The next below is a good starting point for comparing the brightness of a satellite to a star. The Southern Cross is easy to detect, a small but bright constellation and visible all year from Australia, New Zealand, South America etc. The magnitudes are all marked against each star. Do not forget the higher the number, the fainter the object. Mag 6 is the maximum limit for the naked eye.

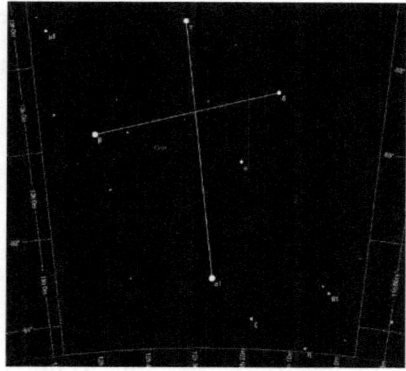

'Alpha Crux' is the brightest star shown at the bottom of the cross at Mag 0.8. Beta is to the left of the cross and is at Mag 1.3 and Delta to the right is Mag 3; a perfect group of stars for comparing satellite brightness.

Chapter 10 Lighting Problems

There are several conditions that limit how many satellites can be seen at any one time. These vary from night to night and place to place and are worth considering in advance to avoid being bored standing in the cold waiting for a moving dot.

Street Lighting: If you stood under a 1kw motion sensor security light that blinds your vision every time you make a move, then you are not going to be very successful in spotting any satellite or stars.

A general orange glow in the sky caused by town lights is Light Pollution. A wet night will enhance this further still by reflected light from the ground upwards. Light Pollution will lower the faintest objects that can be seen. Find the guide stars as suggested in the previous chapter to determine how good the sky is from your position. If you can just see the Mag 3 star Pherkad for example, then just look for satellites down to that brightness only; unless you are more experienced and able to spot them using binoculars – such observing techniques are discussed ahead in this book.

A note for local government planning; an increasing use of pure white LED street lighting is causing a problem for astronomers and wildlife. Most nocturnal forms of wildlife such as hedgehogs, bats & owls can tell the difference between day & night as long the colour temperature of lighting is completely different to daylight. However, white light is the same as daylight and can affect such creatures' habits in a big way. The town of Winslow, Arizona uses a narrow band of Orange LED lighting and solved the problem at no extra cost.

With most forms of orange coloured lighting astronomers can filter out the glow at the telescope or with image processing, but light from pure white LEDs are impossible to deal with. Orange or Yellow LEDs are available. On their defence, some of these white lights are dimmer than the originals they replace and can party make up for the disadvantages.

Some parts of the UK are now having their local street lights turned off after midnight. This is to save resources, improve energy security, reduce atmospheric pollution and expenditure. The biggest reason in the UK is the lowering power supply due to coal station closures; to avoid a panic these other (good) reasons are publicly given. Once midnight arrives, the earth's shadow is directly overhead and no satellites will be seen. The only exceptional time of year will be six weeks either side of June 21st in the Northern hemisphere and six weeks either side of 21st December in the Southern. The shadow point will be away from the zenith enough to allow a few low and all high satellites to be seen.

Moonlight: At least moonlight is very predictable. Do refer to lunar calendar tables to show the phase for your chosen night of observing. It may be visible in the evening as a thin crescent. This would not cause any problem, but a full Moon or close to it will drown out all but the brightest satellites. After full moon, it will rise late, perhaps late enough not to cause a degradation of the night viewing. From three days after full moon, it will not rise until after midnight. The Moon rises approximately 50 minutes later each day.

The following chart shows an example; the good dates for satellite observing are from the 8th to the 14th. Then again from the 27th until the end of the chart. All the other dates will be marred by too much moonlight.

Every month will be different as the cycle lasts 29 days 13 hrs – the average Lunar Month.

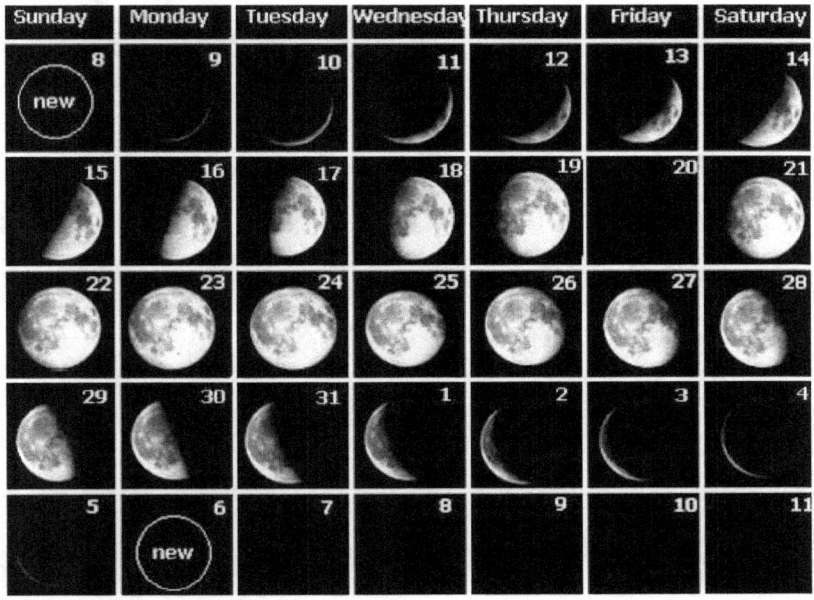

Look up the moon's phases for the month you wish to observe satellites. A little planning can save a lot of frustration. A large Moon in the sky will drown out all but the brightest of satellites. After full moon, it will rise later so as not to spoil the evening sky; approx. 50 minutes later per night. (I don't know what happened on the 20th, perhaps it was cloudy)

www.moonconnection.com

Photography: A large Moon in the sky will also cause serious problems for photographing satellites too. On an exposure of even just 5 seconds, the moonlight will completely fog over your picture. Save your time and effort for another night and just watch the brighter satellites pass instead. Darker skies are required for photography.

Clouds: To solve a cloudy night problem, just get the biggest fan you can find and... just kidding. There is not a lot that can be done about this, but one satellite in particular can still be seen through thin cloud – ISS. It is so bright that if it is due to pass high up, then at least part of the track can be observed. Look around, if a few bright stars that can be seen then you will not be wasting your time completely. Clouds do also reflect city lights and make the sky even less suitable. A few test photos beforehand will give an indication as to the limits of that night.

Even if there are some clouds around, they can make interesting shots and the brighter satellites can still be observed. On exposures of 30 seconds long for instance, moving clouds will be recorded as streaks; a pleasing effect.

Chapter 11   Visual Satellite Phenomena

They are many fascinating events to watch out for. Some of these occur every night, others are quite rare and very unpredictable; it is just a question of luck in witnessing. I have listed examples of these events in a rough order of the most common phenomena to the rarest seen.

Entering and Leaving the Earth's Shadow: As a satellite travels into the earth's shadow, it fades for a few seconds as it enters the Penumbra area then vanishes completely as it passes into the Umbra. The time this takes depends upon the angle of approach. The satellite may head straight to the shadow at 90° and can fade in just 5 seconds. If it approaches from a shallow angle, it could take 30 seconds or more. On rare occasions, it may skim the edge of the Umbra and just make a fainter crossing of the sky without actually fading at all.

The angle of approach can be seen by printing out the sky chart of a satellite crossing from www.heavens-above.com .

Very bright sightings such as ISS or an Iridium satellite flare can appear pink. Imagine you are on the satellite, the sunset will be seen through the earth's atmosphere and red / pink light will get through but not the rest.

Iridium Flares: There was a constellation of satellites known as Iridium. A global communication system included satellite phones. They were at around 11,000 km up and normally hardly seen with the naked eye at Mag 4-7. These had highly reflective gold-coated antenna and reflect sunlight like a mirror. The following picture is of a full size Iridium Satellite model at the Smithsonian Air & Space Museum, Washington DC.

These could often even be observed during the day as a flash of light lasting 3 seconds or more. At night, the flash can be observed for up to around 15 seconds. As they were all in polar orbit, this amazing phenomena could be seen anywhere on Earth and become known as Iridium Flares.

I have recorded an Iridium Flare through an image intensifier…
Look up the webpage on 'Iridium Flares.'

An Iridium Satellite can flare from being virtually invisible to become the brightest object in the sky within a few seconds. For details of how to view them, refer to the chapter called Heavens Above. These can even be seen in broad daylight.

As of December 2018, these wonderful satellites were being brought down one by one over the Pacific Ocean, as they were simply outdated. The last was down by the end of 2019. They are being replaced with the Iridium Next series; these do not flare due to a different design (below). They are sadly missed, but other satellites do produce flares but not often as prominent as these do. *(Next – the replacement Iridium).*

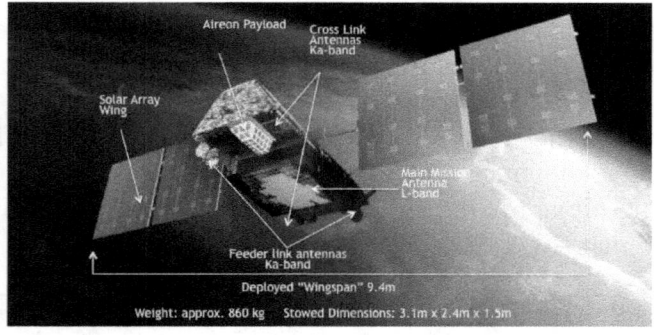

Other satellites however, show a similar phenomenon. The following image is a pair of unknown satellites invisible to my eye, but as they changed orientation to the Sun, for around 10 seconds, they both brightened to an impressive Mag -2 or so.

These were observed from my hometown in September 2023. For the flares to be synchronised, both satellites must have had the same orbit, physical shape and orientation.

Flashing Satellites: There are spinning satellites that are uneven in surface area; elongated along one side. Most such examples are rocket casings that are cast adrift after launch. They are often the final stage of a satellite launch that achieve orbit too. They can appear to flash bright to dim in a couple of seconds or very slowly taking a minute or more to rotate once. Some of these can be regular satellites where control has been lost, thrusters have fired to gain control but a failure of some kind has occurred. However, most are simply discarded rocket boosters or satellite casings.

To record the length of a complete rotation accurately, chose a point in the flash such as the peak brightness, start a stopwatch with a finger, not by looking at it. Count five complete rotations as an example and stop the stopwatch at that peak in brightness. Simply divide the time by five. The longer the duration of timing will give you a more accurate reading. As a satellite lowers in altitude in the sky, it will dim due to increasing distance. Also do take into account that as it recedes, your perspective view is changing and gives a false timing of rotation period; this is the *Synodic Effect*. To avoid both inaccuracies in your observation, limit the timing period to a minute either side of its peak altitude.

The Synodic effect also includes short bright flares that are due to a flat surface on the satellite reflecting the Sun straight to you for a brief moment. It is also not connected to rotation but orientation of the Sun angle.

A slow rotating satellite that faded in and out every 6 seconds or so across the sky. It was around Magnitude 3. I took this in Devon August 2013; dark skies for faint objects.

In other cases, payloads will rotate around an axis (for stabilisation purposes, or when they are out of control after their useful life) and show spectacular reflections of the sun. Polished extensions, such as antennas or solar panels behave periodically as mirrors and cause bright flashes or flares. These can reach negative magnitudes: brighter than any star. Fragments are also taken into orbit in an uncontrolled rotation that can result in short visual flashes.

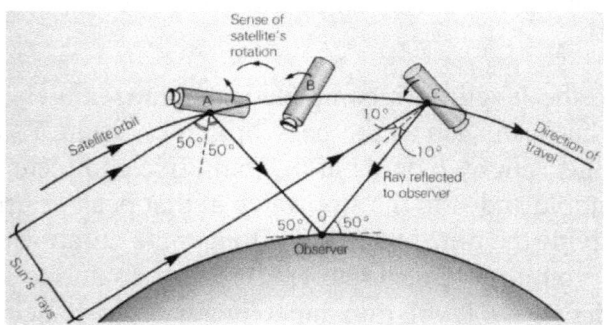

If an elongated satellite rotates, the amount of reflected light to an observer will change and alter the brightness. Some can become invisible to the naked eye then become visible again.

Dockings: You may be lucky enough to witness a docking or undocking event. A second fainter object will then be seen a few degrees away. The next image is such an example of a docking between the ISS and a Dragon supply vehicle.

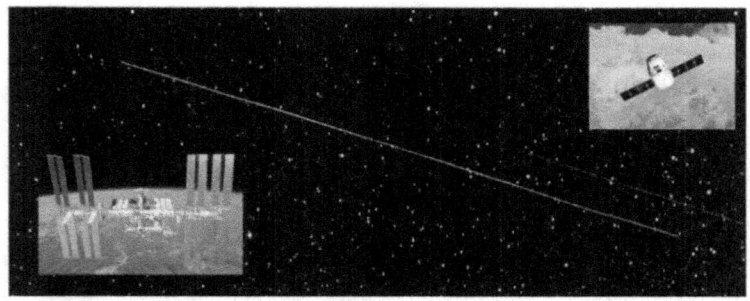

The very first docking I witnessed was in November 1973 when the last Apollo docked with the Skylab space station. As I saw it, both craft were just a degree or so apart. Skylab was about Mag 1 while the Apollo was behind it at about Mag 2.5. The associated Skylab / Apollo 4 capsule is now on display at the Smithsonian Air & Space Museum in Washington DC.

Fresh launch with trailing boosters: Whenever a new satellite is put into orbit, the outer casing or other rocket parts of the satellite often remains close by for a few days or weeks. Just look on the Heavens-Above website (Chapter 29) for pairs of objects that pass over at the same time and altitude in the sky. You should see a few every month; a new satellite is launched every 3 days or so from somewhere in the world.

The following picture was taken in November 2013 by the author. A Chinese military satellite called Yaogan Weixing that is using radar imaging in microwave using Synthetic Aperture Radar or SAR; it was launched just four days before the picture was taken. Shhhh - it is all top secret.

The satellite is the top trail; the lower is the casing. (The camera did move slightly as the exposure began and produced a tiny tail on everything; ignore it please. I lost the infrared remote in the dark that night. Since then I have used a cable version; as it is attached to the camera, they cannot be lost… just do not trip over it or stand on it instead).

The top trail is the Yaogan Weixing satellite; it later tried to search for the missing plane Flight MH370 of April 2014. The Black & White negative process was to enhance the trails more.

Train of Satellites: A group of satellites in a similar orbit, all contributing data following each other is a Train of Satellites.

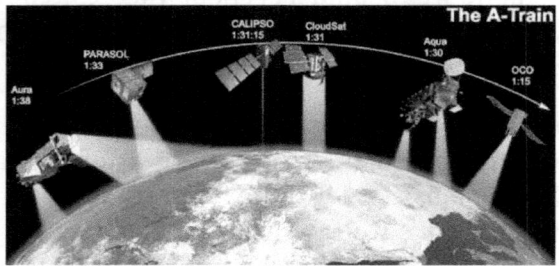

The ultimate Train of satellites has to be the Starlink launches. One rocket casing can include many satellites attached in a row on a mechanical arm or boom. Once in orbit, one satellite is ejected, an hour or so later, another follows in the same orbit. The result is a row of moving dots across the night sky. Over a period of weeks, these spread out across the Earth.

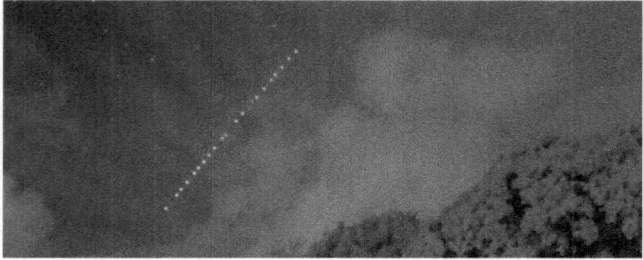

A Starlink Train just 48hrs after launch from the Kennedy Space Center, Florida. Observed from the Isle of Wight; Credit Author.

Formation Flying: Satellites can fly in formation of three in a triangular grouping. This is mainly for live 3D imaging. Military satellites can gain a full 3D view of a battlefield; not just of where ground vehicle and troops are but also of where aircraft of all kinds are too with altitude speed & direction data. Landsat 7 flew as a pair with Earth Observing 1 and produced the first 3D hurricane data that led to new models of storm predictions.

The Swarm satellite formation is in polar orbit between 460km and 520km; all are visible from anywhere on the globe. They are named Alpha, Bravo & Charlie, built in Stevenage UK and launched from Plesetsk Cosmodrome in Russia on the same rocket. This system provides live 3D measurements on the earth's magnetic field caused by the earth's deep interior. This adds to our knowledge of earthquake causes and predictions.

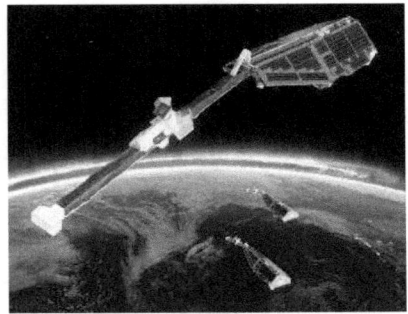

The satellite formation called Swarm. Designed to observe the ever-changing magnetic field of the earth.

Re-entry: All satellite orbits slowly drop in altitude due to atmospheric drag. We usually say that space is a vacuum; no air. The earth's atmosphere just gets thinner and thinner until there are no molecules at all. The vast majority is below 100 km. Even at 36,000 km up at geo-stationary orbit, a few molecules per cubic metre still exist. In time the impacting force, as weak as it is, will slow down the satellite and begin to drop altitude. These machines will remain in orbit though for over 1 million years. The Low Earth Orbit satellites are hitting millions of air molecules as it orbits at 7000 metres per second or so. These have a constant effect on the orbit. For large objects such as ISS, it can drop from 400 km in altitude to 350 km in a year or so. An engine firing boosts it back up. Without such adjustments, it will simply fall out of the sky within a decade. This is precisely what happened to the Skylab space station and entered the atmosphere on its last orbit in 1979. Pieces were found across Australia.

A satellite burning up in the atmosphere, recorded from California; credit Author.

When a satellite is due to fall out of orbit, alerts appear on the Heavens-Above website – check out our chapter *Which is which?* The last orbits will be announced and you may be lucky enough to see it break up into many pieces and witness a spectacular event as it trails with smoke and debris across the sky.

Such predictions are inaccurate at this stage of research. There are many unpredictable variables such as air pressure in the upper atmosphere, the exact orientation of the vehicle which can then change as it begins to re-enter. Once it does, the eyewitness accounts of such events are breath-taking.

The temperatures reached by re-entering debris follow a simple rule… 1m/sec per 1°C. Therefore, a satellite entering at 7000m/sec will rise to 7000°C. This only applies to the exterior. The interior is protected to a certain point. Depending upon the design and materials used, some parts can survive to the ground as demonstrated.

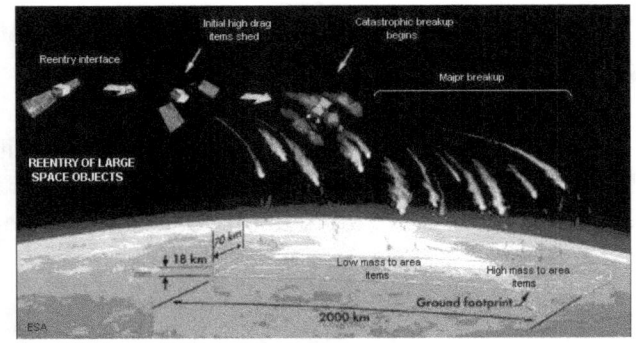

Several clips are on the website; 'Satellite Re-entry' webpage.

Apollo 8 returning to Earth and now seen in the Museum of Science & Industry, Chicago.

Dumping Fluids: When urine, fuels, or other liquids are dumped into space, they disperse into a massive cloud around the spacecraft. If the cloud is big and dense enough, it can be observed from earth.

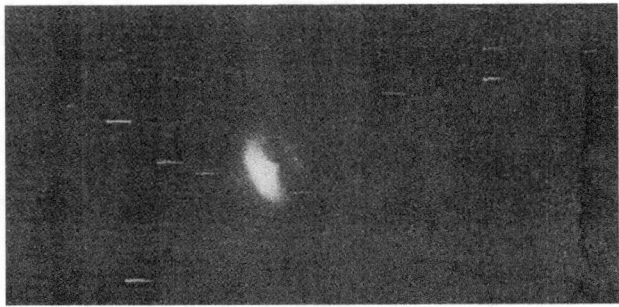

This fuel dump event was recorded from Spain during the Apollo 8 mission of 1968.

Chapter 12 Techniques for Observing

Dark Adaption: It does take a few minutes for anyone to get adapted to the dark. As soon as you step outside, your pupils will widen to allow more light in. The majority of this process takes around 5 minutes depending upon your age. Full dark Adaption takes 45 minutes or so.

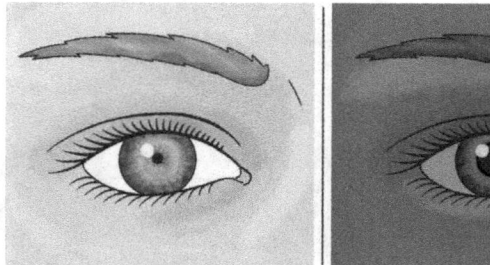

Apparent Width in Degrees: To become familiar with angles in degrees on the sky, the picture below gives a good guide of how to relate to it.

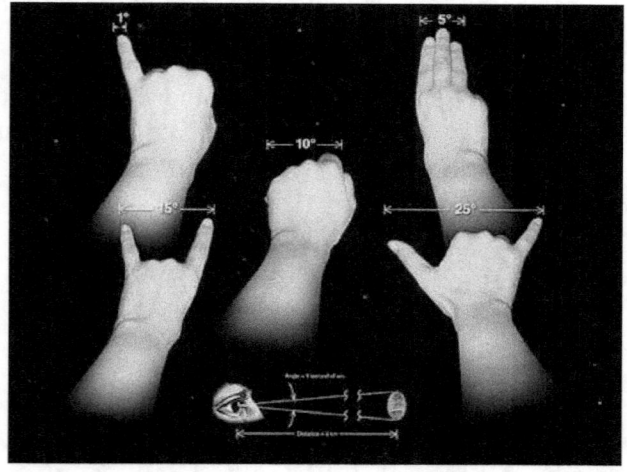

Little finger width; 1°
Two end fingers; 15°
Fist; 10°
Three fingers; 5°
Thumb and little finger; 25°

Naked Eye Scanning: When looking for a slow moving object, keep your eyes fairly still in the general area where you expect to see a particular satellite; a moving object will stand out much faster rather than looking all over the sky randomly.

Binocular Scanning: For satellites below naked eye visibility, move the binoculars in a zigzag fashion along the predicted satellite track working backwards from where the satellite is predicted to be. Scanning at random means, you will miss it; the satellite travelling at 7 km per second or so will not wait for you.

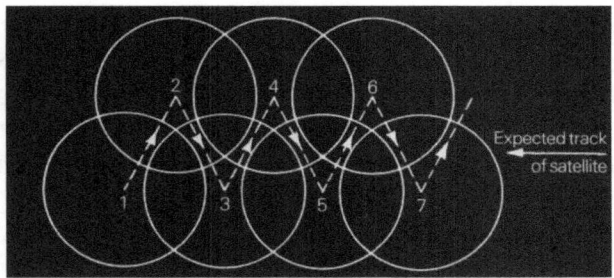

Geo-stationary Satellites: These do not move against the background sky. The Earth rotates minute by minute so the stars slowly change position. Use Heavens-Above (Chapter 26) to find out the compass bearing and altitude above the horizon for a group of these. Such objects can be viewed through large aperture binoculars.

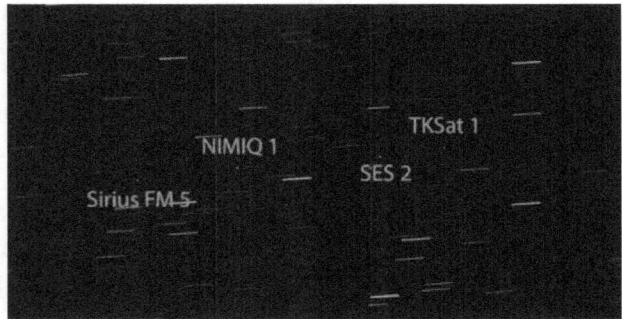

Use a five-minute exposure with a camera aimed at the 'Celestial Equator on the sky (Orion's Belt is a good guide) and Geo-Stationary satellites reveal themselves virtually all night. The stars will streak, but the satellites will not.

Chapter 13   Which satellite is which?

The key to the entire hobby of Satellite Spotting is being able to tell one satellite from another. From the ground we just witness a moving dot. To improve the interest in this object we have to know when they pass over, what it is, when it was launched, who built it and what it is doing. Other matters of interest include; will it go into shadow or is it tumbling to make it flash?

As such information changes by the hour, the only way to obtain this for your hometown is to use the Internet. There are several websites around including NASA for the bright satellites. I have listed a few below but have concentrated on one in particular. Heavens-Above seems to be the most comprehensive site in the world for satellite spotting. It is completely free; you do not have to register in any way and does include thousands of objects bright, faint, live, defunct, junk, low and high orbiting satellites of all nations.

www.n2yo.com A good site to see the live position of various satellites but full of adverts that cut into your experience.

http://spaceweather.com/flybys/country.php This one is good for starters. Not very detailed though, no sky charts, but does include compass bearings and an indication of brightness. No information on Iridium Flares or other unusual phenomena. But yes good if you don't wish to go into too much detail.

http://spotthestation.nasa.gov/sightings/   This one if for the Space Station only.

www.heavens-above.com   This one is the business. The No.1 satellite spotting website for the beginner and experienced observer with full sky charts, compass bearings, brightness information, enter or exit shadow positions. Predictions can be sought for several days ahead or even in the past. You may have seen one in particular the previous night and you are curious to what it was.

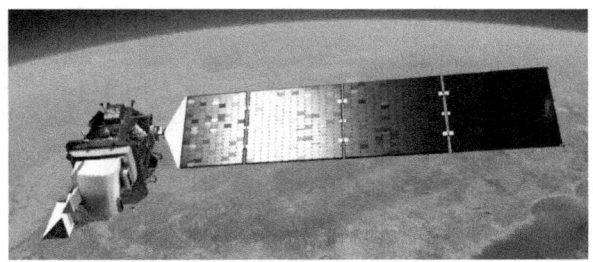

Landsat 8; knowing which satellite is passing over together with its function bring home the reality of how our modern civilisation operates.

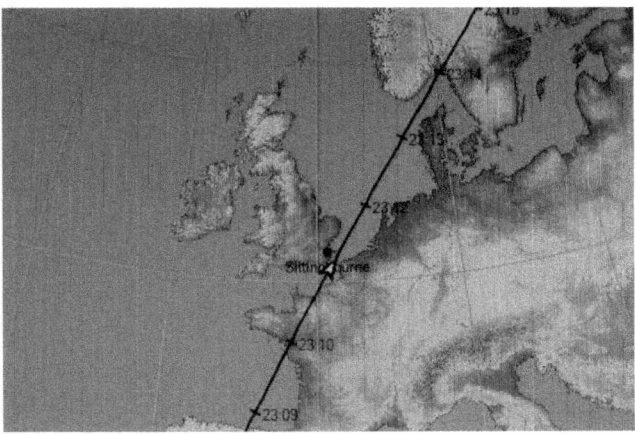

Heavens Above: The Heavens Above website really is the key to the hobby of Satellite Spotting; the World's most complete site available to the public for free. You do not need to register and is easy to use following the step-by-step guide in this chapter. Use the link below after reading this chapter and familiarise yourself with the site. www.heavens-above.com

All the satellite predictions are up-to-date. The further ahead you look in time though, the less accurate the forecast. If you intend to observe tonight or tomorrow; then there is no problem. If you want predictions four weeks ahead, then print it out, but reprint the same night's forecast on the day if possible. Satellites can change orbit by firing a thrusters, or through atmospheric drag if in a low orbit. One of them may have even been deliberately de-orbited and will never see it again.

If you saw a satellite the previous night, then the same information is available. Change the date on the predictions list and look up the time, direction and rough altitude in degrees above the horizon when it reached maximum height. The closest match should be the one you witnessed.

To become familiar with this website at a rapid pace I suggest the following path;

Register: Give a few details to Heavens Above, then you do not have to reset your location every time; just log in with a user name and password of your choice. You would also have the opportunity to post your own observations if you wish on unusual events or ask general questions to other users. You will not be bombarded with emails and adverts at any time.

Setting Location: After finding <u>www.heavens-above.com</u> the site starts assuming you are at zero longitude and latitude. Click on Change your observing location…

Configuration
  Login (optional)
  Change your observing location
Satellites
    10-day predictions for satellites of special interest
      ISS
      Tiangong 1
      N. Korean satellite
      X-37B
      HST
      Envisat
    Satellite database
    Daily predictions for brighter satellites
    Iridium Flares

… Then type in your hometown in the top line and click search. Once you see your hometown accepted with a map, go to the bottom of the page.

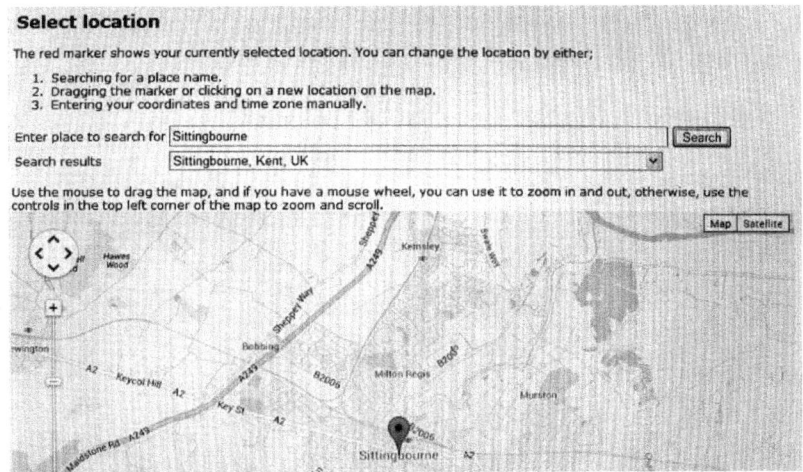

At the bottom of the page, you will find the 'Update' button. Click on it and the site returns to the home page, but from now on all the predictions are from your location. Regarding time zones and Summer Time / Daylight Savings, this is already taken into account. All the times for predictions are your own local time to avoid confusion.

Choose an option: From this point, you have the option of looking at a simple list of satellites visible that night. Click on the button as shown and a list of 50 – 150 satellites will appear as potential targets for you. The highest numbers will occur around March and September due to the Earth shadow angle and length of the night.

Satellites

10-day predictions for satellites of special interest
ISS
Tiangong 1
N. Korean satellite
X-37B
HST
Envisat
Satellite database
Daily predictions for brighter satellites
Iridium Flares
Spacecraft escaping the Solar System
Amateur Radio Satellites - All Passes
Height of the ISS

This list could display over 200 satellites, so to reduce it, click on 'Minimum Brightness 3.5 and click on 'Update.'

Daily predictions for brighter satellites

Month [May 2014 ▾] Day [17 ▾]   ○ Morning   [Update]   [Reset to now]
                                 ⊙ Evening

Minimum brightness: ○ 3.0   ◉ 3.5   ○ 4.0   ○ 4.5   ○ 5.0

Satellite	Brightness	Start			Highest point		
	(mag)	Time	Altitude	Azimuth	Time	Altitude	Azi
Meteor 1-15 Rocket	4.4	21:18:04	10°	N	21:23:52	78°	E
H-2A R/B	2.6	21:19:37	10°	NNE	21:23:47	58°	E
H-2A R/B	3.0	21:22:17	10°	SSE	21:26:19	81°	W
SPOT 1/Viking Rocket	4.1	21:22:51	10°	S	21:27:58	61°	W
Cosmos 1939 Rocket	3.0	21:23:05	10°	SSE	21:27:16	88°	E
Gbstr 26 Del rocket	4.2	21:23:08	10°	W	21:29:13	84°	S
Cosmos 1943 Rocket	3.7	21:24:54	10°	SE	21:28:14	16°	

If you wish to change the date, the option is above; but not explore to far into the future as the predictions become less reliable.

The easier satellites to observe are those that fly almost overhead. Study the list of 'Highest Point' figures; the closer to 90° the better as this is overhead (the Zenith). The lowest I normally try to observe around 35° high. I chose one example next – **Cosmos 1943**, launched in 1988. It was predicted to be Mag 2.1 and pass at 87° above the horizon at 23:11pm.

CARTOSAT-1	4.0	23:05:15	21°	S	23:07:56	55°	W	23:12:12	10°	NNW
Cosmos 1656 Rocket	3.8	23:05:36	20°	S	23:09:15	54°	ESE	23:14:37	10°	NE
Cosmos 1943 Rocket	2.1	23:05:37	10°	SSW	23:11:11	87°	ESE	23:16:45	10°	NNE
ASTRO F (AKARI)	4.4	23:08:41	35°	SSE	23:10:02	81°	ENE	23:13:22	10°	NNW
CZ-4C R/B	3.9	23:09:57	23°	SSW	23:12:03	47°	W	23:15:38	10°	NNW

After clicking on the time, a full sky chart will appear with the satellite track marked on it with direction of travel and times.

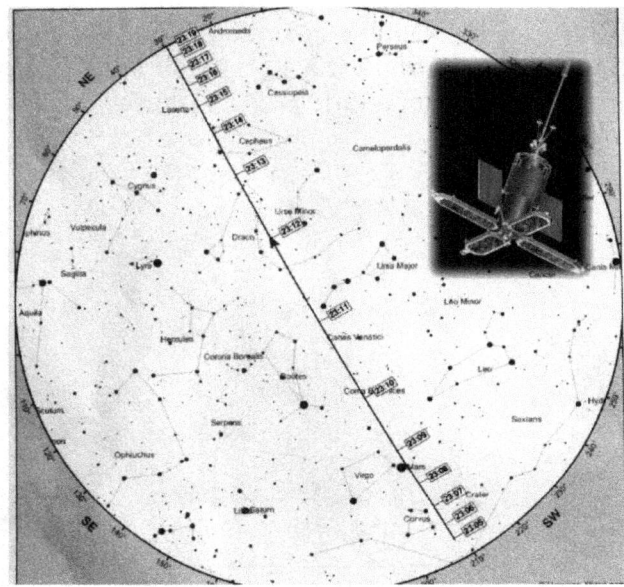

Go outside 10 minutes before the predicted maximum altitude of **23:11** in this case and get dark-adapted. Look for the overhead stars and see if you recognise them. It does not matter too much if don't, the moving satellite should be seen anyway regardless. In this particular case, the satellite was a rotating cylinder. This means that the observed brightness may be completely different to prediction. Brightness of an operational satellite is accurate.

To increase level of interest and home to you what you are actually seeing, click on the satellite name itself and details of the launch etc. will be shown.

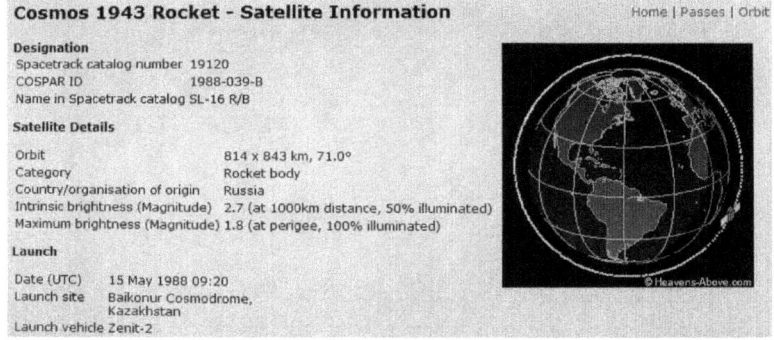

Cosmos 1943 Rocket - Satellite Information   Home | Passes | Orbit

Designation
Spacetrack catalog number 19120
COSPAR ID   1988-039-B
Name in Spacetrack catalog SL-16 R/B

Satellite Details

Orbit   814 x 843 km, 71.0°
Category   Rocket body
Country/organisation of origin   Russia
Intrinsic brightness (Magnitude) 2.7 (at 1000km distance, 50% illuminated)
Maximum brightness (Magnitude) 1.8 (at perigee, 100% illuminated)

Launch

Date (UTC)   15 May 1988 09:20
Launch site   Baikonur Cosmodrome,
                Kazakhstan
Launch vehicle Zenit-2

If you wish to see the satellite again, click on 'Passes' or more detail of the orbit use 'Orbit', it's a very well laid out website as you can see already. After a little practice, you will not need this guide any further.

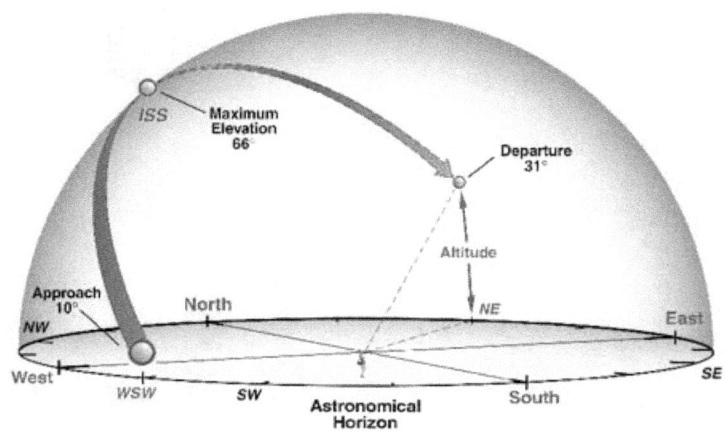

Above; if you don't wish to learn the constellations, then all you really need are compass bearings and an understanding of elevation in degrees. 0° is the horizon, 45° is half way up and 90° is overhead. All these details are on Heavens Above.

Next is a close up of the actual event I observed. As the object was spinning, the amount of surface reflecting sunlight was changing second by second. It was invisible to my eye to start with then brightened to around Mag 3, then faded again. It repeated itself every 20 seconds or so. This object can now be classed as a 'flashing satellite.'

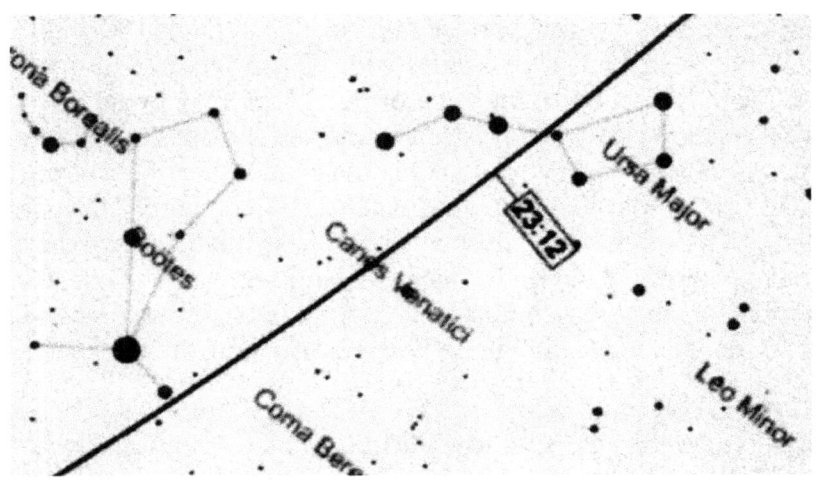

The predicted path of H-2A; US rocket booster 19th May 2014.

The next photo is of the actual satellite H-2A. The track cut short due to a 15-second time exposure. A longer exposure under the conditions that night would have fogged out the picture as there was a little heat mist reflecting local streetlights. Compare this against the predicted track above.

Iridium Flares: Witnessing these is a hobby in itself. Three gold-coated antenna on each Iridium satellite reflect the Sun almost perfectly down onto the ground. Their orientation to within a couple of degrees is also published as well as the orbital details. With both data sets combined, Heavens-Above is able to pass on prediction details of the reflected sunlight.

As the antenna are not perfect mirrors, the reflected sunlight sprays out to a beam several miles wide. The brightness falls off away from the centre of the cone of light. You may be lucky to have the centre pass right over you at Mag-8 (The brightest planet is Venus at Mag-4). If you are not at the centre, then the website provides a simple map showing where it is compared to your position. Travel there in advance and witness the full flare, they can even be seen in daylight. Once you have seen such a flare it's easy to become obsessed into seeing more. Each one last around 5 – 20 seconds depending upon how far you are from the centre of the track.

To get started just find the 'Iridium Flare' option under 'Satellites' on the website after you have input your location. A list will be shown of all the possible sightings for several days in advance. Just change the date option at the top. As of Dec 2019, there are sadly just two such satellites left to observe.

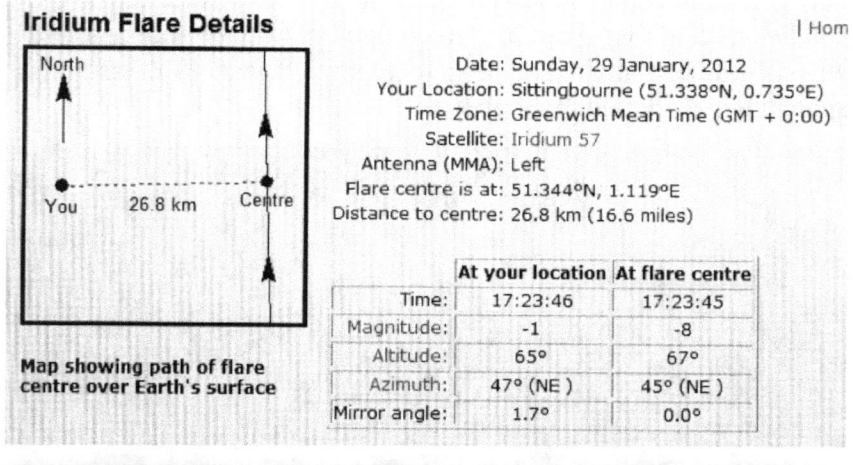

Iridium Flare Details | Hom

North

You   26.8 km   Centre

**Map showing path of flare
centre over Earth's surface**

Date: Sunday, 29 January, 2012
Your Location: Sittingbourne (51.338°N, 0.735°E)
Time Zone: Greenwich Mean Time (GMT + 0:00)
Satellite: Iridium 57
Antenna (MMA): Left
Flare centre is at: 51.344°N, 1.119°E
Distance to centre: 26.8 km (16.6 miles)

	At your location	At flare centre
Time:	17:23:46	17:23:45
Magnitude:	-1	-8
Altitude:	65°	67°
Azimuth:	47° (NE)	45° (NE)
Mirror angle:	1.7°	0.0°

Chapter 14   Satellite Track Photography

The simplest form of recording satellites is to photograph them crossing the night sky. With exposures shorter that 50 seconds, the stars remain virtually as dots while the rapid moving satellite will trail within the field of view.

My very first picture was taken in 1977 of the then abandoned US Skylab space station. The Daily Mail newspaper in the 70's used to show the Moon's phase, sunrise & sunset time on page two. Whenever a bright satellite was visible, the time of the crossing would be included too. The exposure was around five minutes long so the stars have trailed due to the Earth's rotation.

This was taken on an old Russian Zenith E camera with 400 ISO B&W HP5 film. I did not have a tripod at the time, so I laid it back on Mum's trelliswork for her roses and pointed the camera straight up. It was just steady enough without wind to buffet it during the exposure. To focus such a camera on infinity was very easy; just rotate the focus ring as far as it will go anti-clockwise. Digital is not always so easy, but other advantages far outweigh the disadvantages. (I did knock off a rose, do not tell Mum).

The US Skylab space station as it was in 1973 after some repairs by the first crew- NASA

Settings Required: It is important to be very familiar with a camera for any night sky photography. Do use the instruction book rather than guessing as I have done many times. We shall assume that all cameras used for night photography are now digital.

Regarding memory cards, the 45MB/s should be the minimum. This reduces battery consumption and wasted time between images.

With any Digital SLR camera; the following settings must be understood and experimented with.

Field of View (FoV): Most cameras come with a Standard Lens; or at least a lens that can vary from wide angle, through a standard lens view to a slight zoom. This would be preferable. A wide-angle setting will obviously allow a larger part of the satellite trail to be recorded. In addition, as the Earth spins, the stars will trail on the picture and become short streaks, but on a wide-angle setting, this effect is minimised. On exposures of less than 40 seconds, the stars should remain as points of light rather than streaks. Only the moving satellite will produce the trail you want. If you zoom in to the sky, then star trails will be enhanced, shorter exposures will be needed; 10 - 20 seconds.

ISO (International Standards Organisation – was known as ASA) This is a measure of how sensitive the camera is to light and had its origins when film was used. The higher the number, the more sensitive it is, the fainter objects will be recorded but this does include light pollution to. When astrophotography was based on film, the most common ISO available was 400. Kodak that was set at 1600 produced a specialist film. A reasonable mid-range priced digital camera often goes up to 12,000 ISO; perfect for night sky photography. The top range of cameras can now go up to around 400,000. See the advantage so far? (*The Sony Alpha 7S & 9S can go up to 409,600 ISO*).

Refer to the Manual to set ISO. If its set too sensitive and you have some light pollution then your image may fog over with bright orange glow and ruin the picture. Experiments by photographing the stars before the satellite crossing would be very helpful. ISO 400 is the minimum considered for night photography. ISO 1600 is a safe setting for an average sky and will record a moving satellite down to Mag 4 or so. ISO 3200 will record fainter satellites down to around Mag 5. Ideal if you have a dark sky and little or no moonlight.

You can try higher settings, but these only work generally with very dark skies where the Milky Way can be seen. A series of experiments on the night will help you decide. Try each ISO setting and expose for say 30 seconds each time. Look at the results on the LCD screen. After a little experience, you should be able to judge yourself what ISO setting you can get away with under certain conditions. My camera, Canon 600D & the 1300D can be set as high as ISO 12,800. The image does get 'noisy' though with lots of background rubbish appearing but the stars and satellites do become imaged incredibly fast. Such images can be cleaned up with a little processing.

The next table can be used as a guide for minimum required setting can be used for moving satellites. As the sensitivity increases (ISO) so does picking up stray unwanted light from light pollution, mist, moonlight etc. The higher ISO settings require darker skies. This table is a guide only. The slower moving satellites (at higher altitudes) may be imaged more easily as the light will build up quicker on the chip.

ISO Setting (used to be known as ASA)	Faintest Satellite that can be recorded in Magnitude
200	+1
400	+2
800	+3
1600	+4
3200	+5
6400	+6
12,800	+7
25,600	+8

Stars and satellites are obviously much fainter than objects seen in daylight, hence the longer exposure and higher ISO required. This is precisely why stars do not appear in photographs taken on the Moon or in orbit. The astronauts were in daylight; the stars in space will never appear in a picture under these conditions. Your own experiments will prove the same point; Moon hoax conspiracy supporters never bother.

Aperture: You may have come across the term *f* stop in photography or your camera manual. This is a description of the amount of light that enters the lens compared to what is received at the back. There is a mechanical restriction in the lens itself that alters wider or narrower according to the *f-stop* setting. This is equivalent to the pupil in your eye widening under dark conditions or narrowing under bright sunlight. You should be able to manually adjust this – the camera lens that is. The lower the *f* number, the wider the opening. All lenses are different, just adjust it to the widest aperture i.e.; lowest number that the lens gives to allow the maximum amount of light in.

Ideally set the camera on *f 5, f4* or *f2.8 / the lowest f-stop it has*. If you accidentally take a night picture at *f22* then nothing will appear as so little light will have reached the chip.

Focusing: Discover if your camera has an automatic Infinity Focus setting; if not set the body in manual focus. Rotate the focus part of the lens on a distant object such as a roof or tree at night using the tripod to ensure no blurring. (You can do this during the day if it's easier, but once in focus, it's not always easy to keep the lens in precisely that position for several hours until you are ready for night shots).

If your LCD screen is bright and good enough to show a few stars, then zoom into the screen and focus manually.

If you cannot see the stars on the screen well enough to focus, try using a bright laser pen – green or blue are the best. Red will not work well. Fix a laser pen on top of the camera with tape. Set the camera on manual focus, fire a laser beam on a distant object such as a tree or roof or anything over 30 metres away (not aircraft).

You should be able to see the laser dot far more easily than stars. Focus the camera on that dot. Leave that part of the lens alone. It is now set for the stars. Zooming in and out on the stars should not affect the focus. If your camera has auto focus only, then it may not suitable.

This is the ultimate form of focussing. I tried the technique across the Grand Canyon in 2015 and yes the star pictures after came out in perfect focus. This image was taken under moonlight; the colours came out just as good as daylight photographs, but not seen visually due to faintness.

Take a picture with say a 5-second exposure. Zoom in on the LCD screen to see if it is in focus. Most cameras offer a zoom in and out button to examine the picture more closely. Once it is in focus, try another 10-second exposure on some bright stars. Again, zoom in on the result to see if the stars are points rather than blurry round spheres. Once they are sharp, never touch the lens setting or you will have to start over. This experiment will become second nature to you after a few nights and will get quicker at it. This will ensure a rewarding night of satellite photography rather than ending up with blurred satellite trails that are useless.

If you are purchasing a Digital SLR for Astrophotography, ensure it has a Manual / Infinity focus setting. Many cameras do not, but this technique is the simplest solution. Manual override focus is the number one priority to look for as well as long exposure facility; 30 seconds and beyond (Bulb is for as long an exposure as you wish).

*If once displayed on a computer screen, images are still slightly out of focus (as shown here) it is possible to re-sharpen it by using software such as **Avanquest – InPixio Photo Focus**. It's not expensive and can save many photos; not just astronomical shots. But laser assisted focussing will avoid this problem.*

Shot Alignment: The stars do not show very well in the LCD screen when night photography is required. A green laser mounted on the camera flash unit will automatically be roughly aligned to the centre of view. Use gaffer tape to hold it in position. This can now be fired into the night sky and used as a guide to where the camera is pointing. Once in the correct position for your approaching satellite, turn off the laser. I have left in on in the next picture as a demonstration.

Stabiliser / Shake Reduction: There is an internal mirror that flicks up out of the way of the recording chip every time a picture is taken, this shakes the camera. Many digital cameras have a shake reduction facility. Some astro-photographers advise turning this off as the camera is on a rigid tripod. This saves power from the battery and also reduces time to process after each image is taken. It can take around 20 seconds for the camera to prepare for the next image. By turning off anti-shake facility, this can be reduced to around 10 seconds depending upon the camera model.

After trying out seven different DSLR cameras, it may be best to leave the shake reduction facility on. The camera will use extra battery power and take longer to process and get ready for the next shot, but the improved results is worth the sacrifice. One of the few exceptions is the Sony Alpha 7S or 9S which have no moving mirror at all; the perfect camera for satellite photography. Its LCD screen takes the image from the chip and displays that rather than direct from the lens. There is a slight processing delay to the LCD screen though. As long as the camera is never used for fast action photography such as sporting events, then it will not present a problem.

Exposure time: A typical satellite will cross the sky in around 5 minutes. The standard lens itself does not cover the whole sky so a satellite could travel from one side of the image to the other in around 1min 30 seconds depending upon the altitude above the horizon and the lens used. Therefore, a 1min 50 second exposure should capture a full trail right across the entire picture if you begin the exposure just before the satellite enters the cameras' view. Practice will make perfect as with any skill. If you know where the satellite will cross, then align the camera with a bright star on the leading edge of the view, when the satellite gets close, begin the exposure, or even use the green laser as a marker.

As the Earth spins, the stars will rotate around the sky. They can create a pleasing effect if the exposure is long enough. Only the very darkest skies will allow this to work. The next image is a three-minute exposure and the stars have trailed a little. A meteor crossed the field of view during this time by coincidence. Light pollution from a nearby town is also building up; a little will not spoil the shot too much.

If the sky isn't that dark, then play it safe and keep the exposure to no more than 30 seconds and the ISO setting no higher than 1600. You will capture at least part of the crossing and numerous stars but not too much of the light pollution.

Remote Triggers: To avoid a jerky movement on the camera as an exposure commences, it is best to use a remote trigger device. If you do not have one, just put a black cloth gently over the lens; begin the exposure and pull away the cloth (often known as 'the black hat trick'). A two-second delay in exposing the sensor to light will allow the vibration to die down before full exposure to light commences.

Remote triggers come in two forms; infrared remote and a hard-wired cable with hand button; the latter tends to be more reliable for astrophotography as they do not require batteries that are affected by the cold or damp conditions. The camera can be set on say 30 seconds; by pressing the button on the remote, you commence the exposure without jerking the camera. Gently let go the remote, put your hands in your pockets, and wait. If you set the camera to 'Bulb' exposure time, this is a full manual timer. When ready for the picture, just press the button on the remote; as long as you have it pressed, the camera will continue to expose. Wear gloves if it is cold.

A cable trigger is recommended over an infrared trigger as you will never lose this in the dark and does not require an extra battery that can fail without notice. Many batteries hate cold temperatures. Tie up the excess cable to avoid it trailing on the ground or becoming entangled with the tripod legs. Satellites will not wait for you to sort yourself out before a scheduled appearance. £8 ($10) should do it.

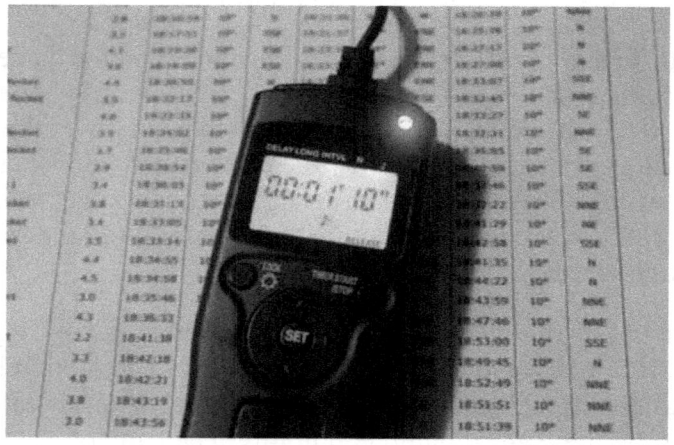

If you want to be flash, you may wish to opt for a Digital Remote These are only around £5 extra than full manual. This will enable you to time each exposure as long as you wish and keep your hands in your pockets rather than keeping a finger on the button. These do take batteries and require a little programming; refer to the supplied manual and do not guess as to how it works. I still prefer to full manual remote for simplicity.

Tripod: A quality tripod is essential. A used tripod should not be any more than £15 ($22) as a rough guide. Do not use tiny floppy weak tripods; use something that is designed for a heavy camera. The weak versions will slowly 'droop' during the exposure and render your starry pictures useless; all that time planning, experiencing perhaps freezing cold conditions... wasted. A video dedicated tripod is ideal. The two main axis are dampened to reduce camera shake further when moving the camera to follow a satellite.

Essential equipment for satellite trail photography include a Digital SLR with standard or wide angle lens, tripod, spare batteries and a remote or black cloth to place over the lens until ready for picture taking. If you have not a remote trigger, place the cloth over the lens, start the 20 second exposure or so then gently whip off the cover.

Batteries: Ensure the battery in the camera is fully charged before the evening (or morning) of astrophotography. Long exposures use up power very rapidly. The cold also has an effect and weakens the battery power output. It is strongly advisable to purchase extra batteries as spares; most are no more than around £5 ($7) on ebay. I usually have four batteries on standby in my pocket. Once they are flat, keep them separate from the charged ones in a different pocket... not rocket science I know. Working in the dark, trying to capture satellites at a set time and position, require exacting practice. They will not wait for you while you change a battery after accidentally putting a flat one back in.

First Trial Run
If this is your very first go at astrophotography, try photographing a plane passing overhead at night. It will trail just like a satellite. When you see the picture on your computer screen, the stars should be pinpoints and the plane should show a nice trail with all its flashing lights spaced out.

After an initial satisfying 'hooray' zoom in and see if it is still in focus. If slightly out, it is possible to re-sharpen it by using some software such as **Avanquest – InPixio Photo Focus.** It is not expensive and can save many photos, not just astronomical shots. The latest version of Photoshop may include the same facility

The previous image is an example of a tumbling satellite seen from Arizona, Aug 2014; a piece of debris rotating every 30

seconds or so. This picture recorded one complete rotation. As the object became sideways on, it reflected very little light, then as it rotated full on, the light increased. The slight 'jaggy' line is due to the point source of light moving from one row of pixels to the next on the chip. It's not the satellite waving around.

This is not a meteor but is ISS passing from right to left into the earth's shadow; October 2014. Picture by a friend, Tony Rickwood of Ullapool, Scotland.

Timings: After gaining some experience at satellite spotting, you may wish to observe satellites that are under scrutiny, such as newly launched spy satellites or an object that is about to burn up. A timed photograph can be used to determine the latest orbit.

All that is required firstly is an exact location of your site. Latitude & Longitude can be looked up on Google Earth. Enter the address and look at the map and details on your screen. Alternatively use a SatNav (GPS / Street pilot) for the same information and look under 'your position' mode.

You will also need to have with you a battery clock or watch that is updated by the **radio time signal** and display the seconds. It must have a push button light for the display. Once the satellite is within the field of view of your camera start the exposure and a mental note of the time to the second and make the exposure last for a round figure of say 10, 15, or 20 seconds. The time of the end of exposure just need to be added.

For instance, 8:10:15pm plus 15 second exposure. Obviously, the completion time will be 8:10:30pm.

Write down both figures along with the satellite name and date before you forget. The entire streak of light must be within the photo edges so the start and finishing points are associated with the two timings and mention which way the satellite was moving on the picture... right to left for example.

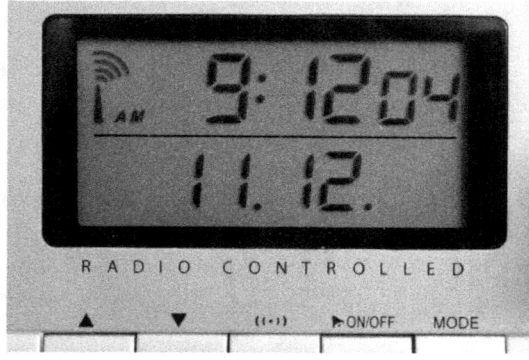

At least two stars in the picture must be named too. Once sent in to Heavens-Above for instance, an expert can use this image and timings to help determine the orbit. Several other such observations will be required from observers around the globe of the same object.

This information is essential for keeping track of satellites about to re-enter the atmosphere. A determination can be made of whether it could be a safety threat. More importantly, attention needs to be drawn to rogue satellites that change orbit for reasons of secrecy.

Simulating apparent speed: Imaging a track across the sky will give you the satellite's path. It can never give an indication of velocity across the sky. However a piece of cloth can. Known as the *'black hat trick'* all that is needed is a dark square piece of material around 40cmx40cm.

The satellite begins to climb into the sky heading toward your position. Commence the camera exposure on 'bulb' setting so the shutter is constantly open. Count to five seconds, and quickly place the cloth over the lens. After another five seconds, whip off the cover and repeat the action another five second later. Continue the process several times as the satellite passes through the camera's field of view.

The result will be a series of dashes and each one represents five seconds worth of flight. The length of each dash will increase, as it gets closer, as its apparent velocity increases.

Once virtually overhead, the dashes will appear evenly spaced out. The result is almost a work of art.

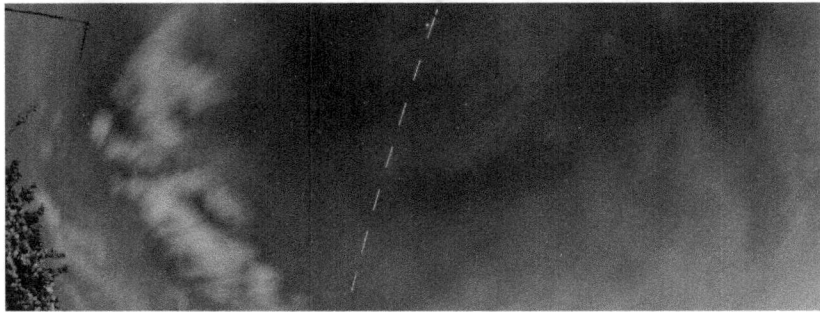

Kit Bag: Once you have produced a standard outfit for satellite track photography, it is good practice to find a bag for the kit. This will save time, improve motivation to go out at a moment's notice and photograph perhaps an unusual event. Within a few minutes of deciding to produce an image, the data is there in the camera and you can upload your picture to a website such as Heavens-Above or Facebook etc. The kit should include;

Camera / Lenses
Fast memory card (Video compatible / 45MB/s second minimum)
Remote Control / or black cloth to cover the lens
Tripod
Red & White LED Torches
Green or Blue Laser
Radio Controlled Clock
Spare Batteries for Camera and Laser + charger
Small Carpet or Rug to kneel on and reduce lost parts
Compass
Notebook and pen

Tip for pleasing result; use the white light torch on a tree or wall etc. within the field of view for a few seconds during the exposure. The object will be lit and the stars and satellite target will not be affected. Practise beforehand to gain the best length of time for illumination. Too much light for too long will burn out that part of the picture instead. This acts as natural frame and adds scale and interest.

Chapter 15   Moving Image Recording

The secret to recording satellites in real-time is obtaining equipment sensitive enough to detect and record them, plus accurate focussing. Many Auto focus systems have a hard time focussing on faint stars: manual override is essential in every case.

Tips: When trying out a new instrument for the first time, practice tracking aircraft during the day. Align the camera so the aircraft is horizontal in the field of view and keep the same orientation as it passes over. Recording a satellite is similar. If the camera is kept horizontal, then its path seems to change as it orbits the earth. Then record bright stars or planets in the sky at night. Play them back on a TV to see how sharp the image is and how shaky. Build up experience on obtaining useful footage and then you will be ready to capture satellite passes that can be shown off on YouTube etc.

Do not be tempted to use high magnification; all that is achieved is greater camera shake. A wider field of view will show more stars for comparison. Try to record rooftops or trees at the beginning or end of a satellite pass for scale.

As satellites move across the entire sky, using any of the following devices may be best hand-held. With practice, a steady image can be obtained. However, it does take time. If a satellite is heading west to east via the South, then keep your feet facing south and rotate your body to the West to begin recording the satellite. Turn your body with the satellite but keep your feet still. In other words, face the centre of the track before the recording starts and turn your upper body with camera in hand to the satellite. Again, practice on recording aircraft during the day.

Digital Single Lens Reflex Cameras (DSLR): Many of these models have video recording facilities with sound. Set the sensitivity to at least ISO 1600 setting first. Use a Video compatible memory card, this ensures rapid recording. Also observe the MB/s transfer rate, 45MB/s should be the minimum; the figures are marked on the card. The camera battery will need to be fully charged too, such recordings take a high toll on battery output power, especially on cold nights.

Camcorders: Some camcorders are ideal for recording the brighter satellite passes. If you are to purchase one specifically for this, then ensure it has zero Lux sensitivity, manual focus and HD quality. If your camera uses memory cards, then only use Video compatible cards and the minimum transfer rate should be at least 45MB/s; I now use 95MB/s.

If the camera has Auto-focus only, then it is not suitable. It will rarely be able to keep constant focus on dots. The lens will just get stuck into recording a blur. I am very suspicious of all these UFO reports taken by a camcorder with auto focus. The points of light always seem to become saucer shaped when the filmmaker zooms in. I get exactly the same effect if I zoom in on Jupiter, or the Space Station etc. I know exactly what I am seeing, but the camcorder will show an apparent flying saucer instead; the multiple elements in the lens system cause the effect. Inexperienced observers swear they have recorded aliens. Manual focus on infinity or a dedicated Infinity only setting is the only way round this.

If you are very patient, set up your camcorder on a tripod and zoom in on the Moon when it is full or a couple days either side. Set the focus on infinity and just record. Some auto-focus cameras will be fine for the moon. Every so often something will fly in front of it; birds, planes and yes satellites. If a phase is chosen say 3-4 days before full moon, it is possible to view it during daylight hours of late afternoon and record satellites transiting it. The same can be said for the Sun at any time during the day; if its projected into a dark box, then satellites will be silhouetted against the solar image.

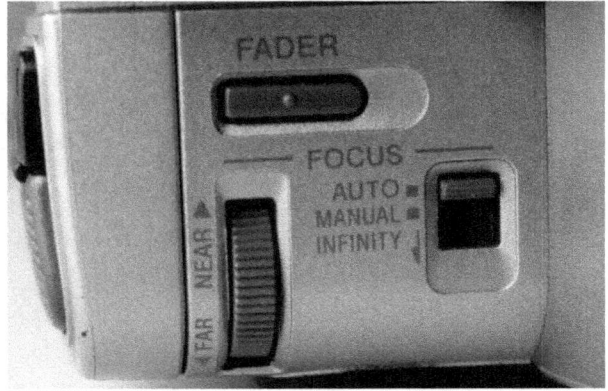

Our Sony camcorder with manual infinity focus

Play recordings back on a TV and wait. There may nothing on it for ages. As each satellite is observed, then that part of the recording can be edited and put aside to add others to it.

Check out our website to see such recordings via YouTube. If viewing direct YouTube, please do ignore any comments regarding 'flying saucers' or 'fake satellites' - they are indeed satellites.

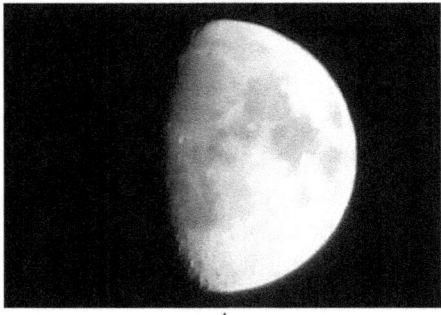

This image was achieved on a HD camcorder with auto-focus. There is a lot of detail on the Moon and very bright for focusing, but recording faint points of light is unrecognisable for auto-focus systems. Seek out manual focus cameras for recording satellites.

Night Vision & Binocular Cameras: The camera shown below was designed and advertised as a day & night wildlife-recording instrument. There are two objective lenses on the front, but only one eyepiece at the back. The other is taken up by a CMOS chip that records fairly high-resolution video in PAL (UK) format; US NTSC version is available. I split the signal with an adapter; one to a monitor, the other to a video deck or DVD recorder.

At night, this instrument becomes incredibly sensitive to light. It can record stars and satellites down to around Magnitude 9, well below naked eye visibility and far beyond any standard camcorder. Most of my own clips on the website have been recorded using this system.

I built the unit above while self-isolating from the Coronavirus. The whole unit is mobile and can be taken anywhere in the garden. The camera is permanently attached; I just turn on the camera and alter recorder to use Line3; the phono-socket. I press record via the remote when ready with the satellite in view. I can commence recording satellites passing overhead that are hundreds of miles up within 45 seconds of set-up. I do not live long enough to waste time faffing around.

The tripod shown is a video-dedicated model. The two main axis for vertical and horizontal is hydraulically dampened to reduce camera shake.

The camera is by Yukon – 5x42mm. The monitor shows a live image from the binocular/camera. This is the best way of ensuring accurate focussing and that all the cables are connected. I have spent many hours recording the night sky only to discover a video connector became detached on the ground or was simply out of focus. A monitor reduces errors. I use a video deck instead of DVD recorder simply because I can obtain them for around £10 ($13) now on ebay. Plus a DVD needs to be finalised before playing on other devices. A new disk has to be inserted every time; a tape does not have this problem. I super scrimp everything; no point in splashing out when you don't have to.

NOTE: With Image Intensifiers, the background sky does have to be dark. If there is a hint of daylight left, then the sky will be lightened to white, no stars or satellites will be recordable.

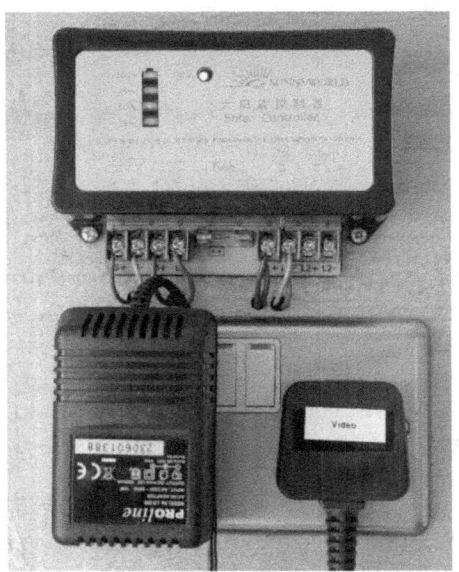

The video deck and monitor are plugged in to a solar power source that includes an inverter to push up the voltage to 230v.

As the system only draws around 40 watts, it is perfect for independent power rather than mains. The batteries in the camera are solar charged too. It is just a personal preference to demonstrate zero-carbon power.

If the number of 'lines' on the camera is the same or less than the recorder, and then no quality will be lost. In the previous example, the wildlife camera produced 400 lines and the Sharp video recorder – 600 lines. So no information is lost as compared to an expensive HD DVD recorder of over 1000 lines; no point if only 400 lines are produced anyway.

To focus this type of camera, two stages are required. First focus on a bright star and observe your monitor to ensure you have the correct set-up. Then refine the focus onto fainter stars as the brighter objects tend overload the recording chip and become a blob. The fainter stars remain as small points so they are easier to home-in on. You are now set for satellites of any brightness.

This Yukon wildlife camera is perfect for recording stars and satellites. It is sensitive enough to record all objects within reach of a standard pair of 7x50 binoculars.

Models of binoculars are now available that have built in digital video recorders. They include a screen and an easy to use menu. The objective lenses are not that large, I am not too sure as to how faint a satellite or star these can record.

NOTE: Do ensure such models are Image Intensifiers rather than just night vision. The latter will not work on the night sky very well.

If money is no object, there are rare binocular models that have detachable eyepieces. One can be used to piggyback a camera instead of looking through. The other eyepiece you will still be able to use for tracking and focusing. These will not be driven to follow satellites or even the stars, but will certainly be enable you to record satellites in real time out to tens of thousands of km. The next pair shown is by Vixen and can be found new or used on ebay for around £2000 ($2500) used as a guide.

These eyepieces are removable for possible camera attachment; visit your local camera store or Astronomy club for advice.

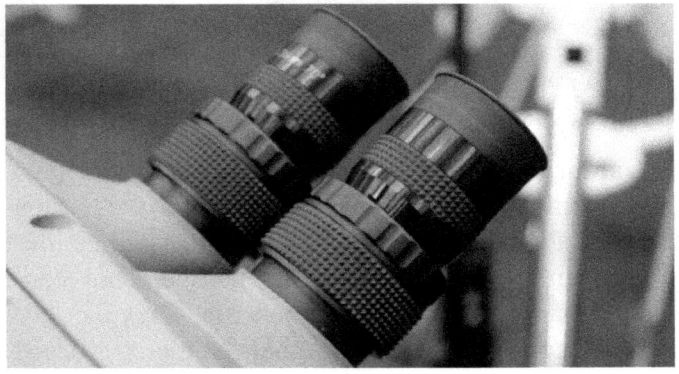

Latest model available, 2021: A company called SiOnyx has produced a range of image intensifiers that are perfect for recording satellites. All four models offer true night sky recording, objects below naked-eye visibility are captured with ease. The four models are known as Aurora Sport, Aurora Black, Aurora Original and Aurora Pro. Prices currently range from £359 to £950 in corresponding order. As of 2023, I now have both the Sport and the Pro.

The two higher end models include a date, time GPS position on your recording, plus a compass bearing. Although these only appear a hidden metadata rather than stamped on the images and videos. All include a microphone to document live commentary if required. Files are stored on Micro SD cards in the standard MOV format: recognised by most platforms including YouTube. The cheapest model, Sport, has all the recording features that the more expensive models has. Only the top of the range Pro version has a better light sensitive chip. I have no doubt that these will become a standard piece of kit. Our films using this camera is posted on www.outerspacebooks.com/my-satellite-clips.html

All Sky Cameras: These come in various forms. The have a very wide-angle lens from 140° to 170° so they are capable of covering virtually the whole sky. Choose the highest definition camera you can afford and the most sensitive. They can record all aircraft passing over, as well as meteors. They can give out a standard PAL or NTSC video signal. Some are designed to record on a laptop. They can be as low as £75 ($100). Fish eye lenses for digital SLRs are perfect for this too but can burn a hole in your pocket. Do investigate many sources for these and get the best deal possible while keeping in mind the image resolution.

A meteor captured with an all sky camera from Huntsville, Alabama.

GoPro Cameras: A range of sport dedicated cameras that can be used for Astrophotography. The GoPro Hero 11 in particular has 'Star Trail' capability. This can be set for taking 30 second long exposures at ISO 1600, sensitive enough to record the brighter satellites. The difference here is that picture after picture is exposed and the camera accumulates each shot into a movie. For satellites to show, select the *'long'* or *'constant'* star trail option. The menu is straightforward but does require time for experimentation. Search for demonstration clips on YouTube.

For star-trail movies, the internal battery will last around two hours. This will give just eight seconds worth of film. Such *time-lapse* recording will speed up the sky rotation and the passing of satellites by 900x. If longer shots are required, consider buying an external battery pack. The can be connected via the camera's mini USB port. I use one with 50,000mAh supply. This can last all night for several nights before recharging again. Time Lapse shots lasting up to sixty seconds long during the winter are now within range.

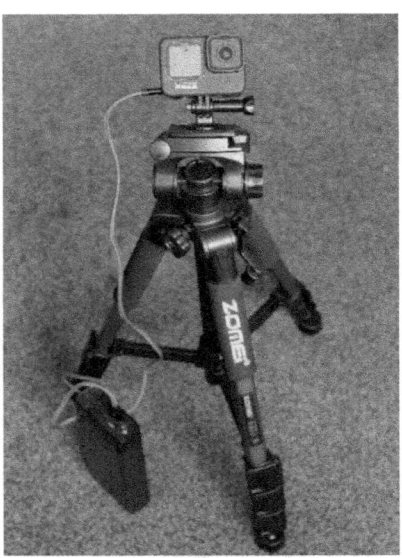

My GoPro Hero 11 with the external power pack.

Chapter 16   Telescopic Imaging

Photographic moving objects through a telescope is incredibly challenging. An astronomical telescope is normally driven by a motor to follow the stars. Satellites move in a different direction and speed compared to the apparent motion of the stars due to the Earth's spin.

Satellites that are very high up, around 10,000 km or more may have a chance of being imaged as they move slowly enough in your field of view to capture them. The example below is of Apollo 8 when it was around 50,000 km from Earth.

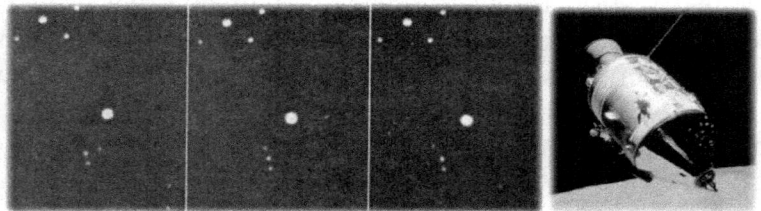

Above; The Apollo 8 Spacecraft on its way to the Moon, 22nd Dec 1968. Photographed by the Yerkes Observatory, Chicago.

Geo-stationary satellites are just that; stationary above the Earth's equator. They remain in the sky as they are rotating at the same speed the Earth is; once every 24hrs. The telescope can now remain perfectly still, no drive working. Use the lowest magnification for a wider field of view to increase the chance of imaging the target and reduce amplified vibration from wind etc. Set it up on the general satellites' position discovered using Heavens-Above and take a 10-second exposure or so. The result should be trailed stars and a satellite or two as a point.

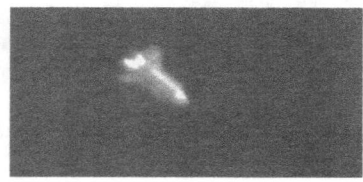

The Space Shuttle Endeavour on STS 57 Mission 23 June 1993.

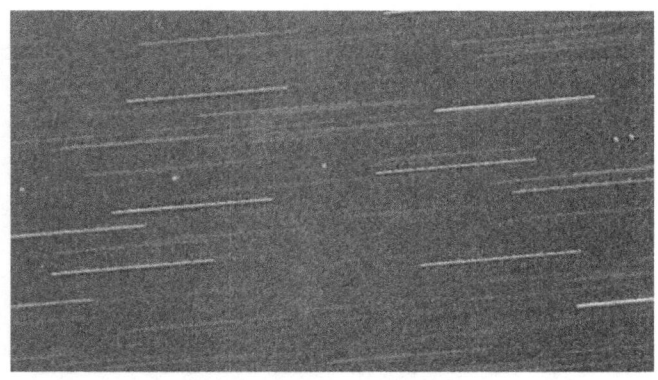

Above; five geo-stationary satellites imaged in 2009 through a low power telescope. These are always positioned around the Celestial Equator in the sky. Check a star map for your calendar month.

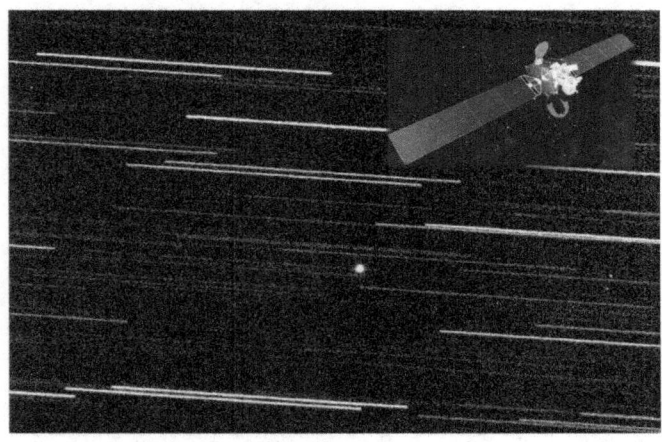

Image of Geo-stationary satellite Intelsat 706; taken through my 5" refractor. It was given a 20 second exposure; the stars trailed across as the Earth rotated, the satellite was a giveaway as it orbits; matching the Earth's speed.

It you have a programmable drive system on the telescope that can slew rapidly across the night sky, it is possible to take close up images of the larger satellites that are less than say 1000 km up. More detail on how to accomplish this will be included in future editions of this book.

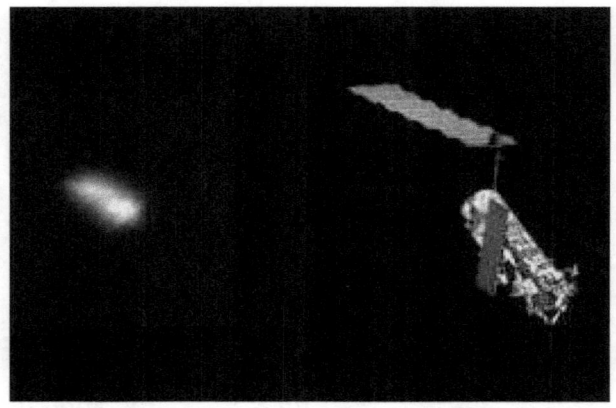

Above; Envisat photographed by Josef Huber of Germany in 2005. The image of Envisat is on the left compared to a model in the same orientation to the right.

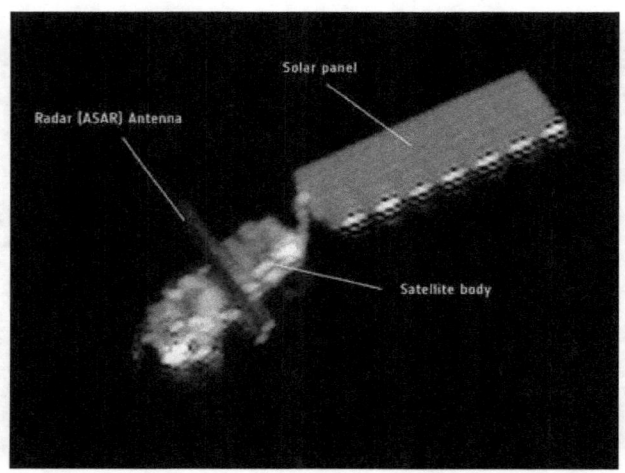

A picture of Envisat taken via another satellite in orbit 2014. Such images are often classified. Credit ESA.

Anyone who wants to investigate driving a telescope to follow a satellite can explore websites such as www.heavenscape.com.

Not all telescopes are compatible. Some Celestron NexStar telescope range are and the Mead Autostar LX200 telescope range certainly is. A few commands can be set using the software provided and the rest is calculated and transferred to the telescope. With the right type of camera attached, some amazing pictures / footage can be captured in real time.

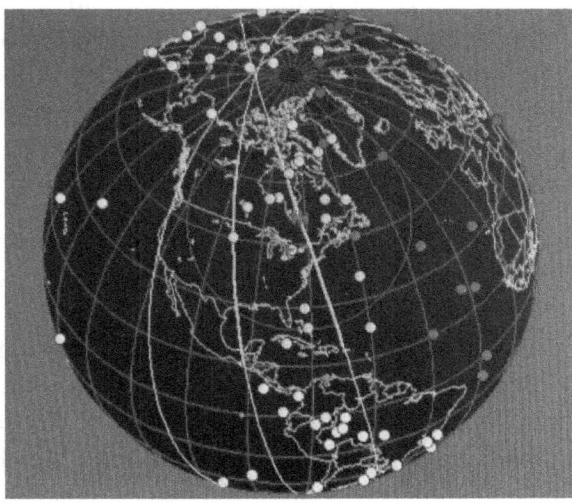

Select from the Autostar online database the satellite you wish to image and the tracking data can be downloaded for your location. The drives on the telescope should follow the satellite.

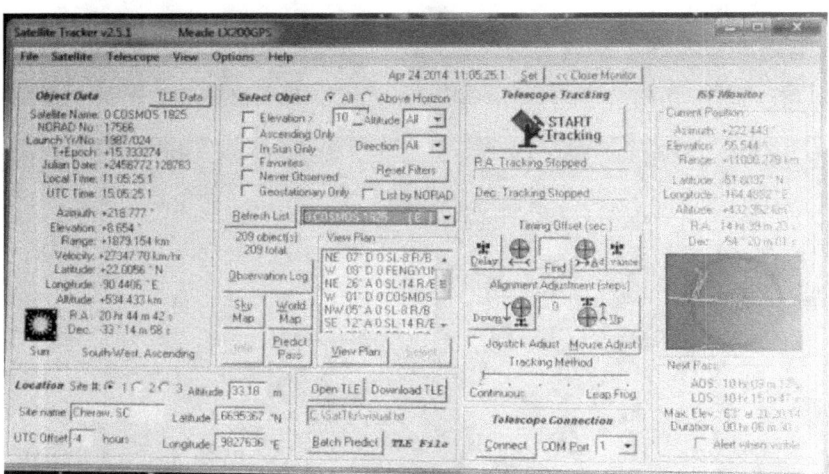

More about this process will be published on the website in the near future. Obviously, a printed book cannot keep up-to-date in the same way.

Manual Tracking: It is possible to track a satellite manually. Strap a green laser onto a telescope and ensure it fires a beam exactly where the telescope is pointing. Alternatively, purchase a telescope finder that has a red central illuminated dot. Use a low magnification eyepiece of around 20-30x. Attach a standard Digital Single Lens Reflex (DSLR) camera to the eyepiece and obtain a focus by testing it on a star or the Moon. (The laser will not harm a satellite or any astronauts).

Wait for any target satellite and aim the laser-guided or red-dot viewer telescope toward it. Begin recording using the video facility on the DSLR. Follow the satellite by using your parallel view finder. Every so often the satellite will be in the field-of-view of the camera. With practice, the amount of time the telescope is spot on the target, will increase. A friend of mine specialises in this and has gained these amazing results...

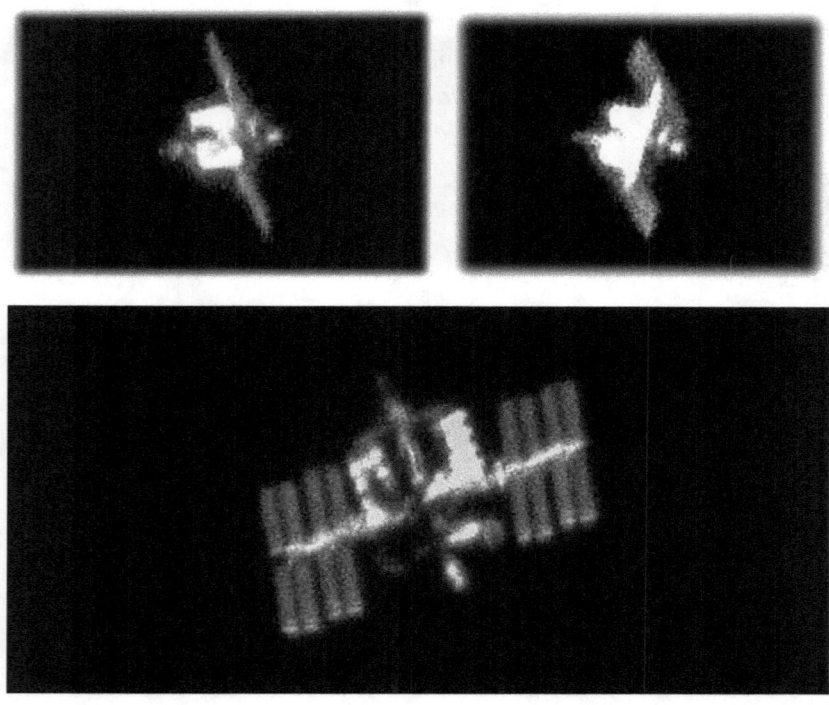

Credit; Art Fentaman March 2021.

Chapter 17   Simple Image Processing

Once you have an image of a satellite passing over, you may be a little disappointed with the picture itself. Perhaps there are not as many stars as you would like or light pollution has fogged a red or green glow across a part or even the whole image. With film photography, there was very little than could be done to improve it, you were stuck with what you had; a little 'pushing and pulling' could be done in the printing process but that was it. With digital, the image can be rescued to a much higher degree. (A white glow from LED street lights cannot be altered, as stars are generally white too).

There are software packages costing hundreds of pounds / dollars. As a super scrimper, this is not my style. Save your money for more important things in life, an acre of land in Arizona costs less. Free versions of advanced programs are sometimes available such as Photoshop PS2 but rare. I use Photoshop 7 that can be purchased on a CD ROM on ebay for around £5 ($7.00). I find this more than adequate for touching up starry photographs. Other free or low priced image processing packages are around.

This book is aimed at the beginner; people that perhaps have never taken any pictures of the stars and satellites before. Therefore, I will not bore you with processes such as 'Dark Frames' and 'Layers.' These are not really required to produce pleasing images of satellite crossings. They are for the perfectionist; later versions of this book will deal with this area.

There are some processes that can be applied to scanned images of film that reduce scratches or grain; these should never be applied on starry images. The software will think the stars and satellite trails are scratches or dead pixels etc. and so 'intelligently' remove your entire image. There are only a small number of key features that need to be learned to improve your satellite trail images.

Example: I have chosen one image to improve, although it does not include a satellite. The original picture does not show many stars and light pollution has altered the true colour of the sky. You may recognise the constellation Orion and Jupiter is seen to the right.

Step one is to change the overall colour of this yucky green. Use 'colour balance' and play with the slider controls until you find a more pleasing and natural colour. You can try clicking on 'Auto Colour' instead but it can go too far. If you do not like the result just click 'undo' or 'cancel.'

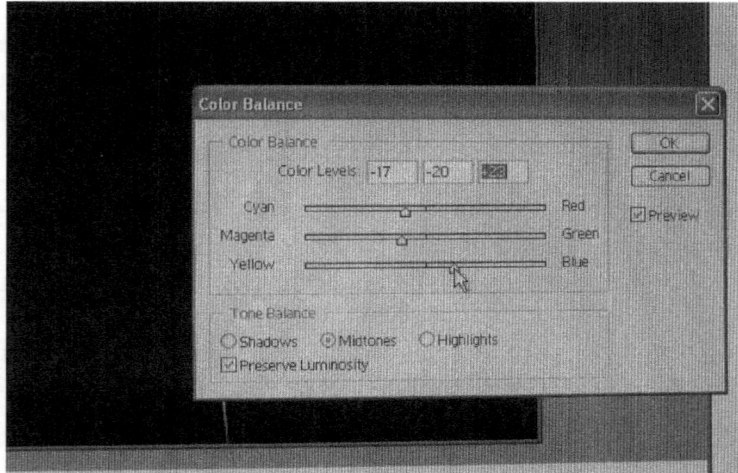

The next step is to improve the number of stars in the image. The CMOS chip almost certainly captured some data of stars that don't show on your image. Using a facility called 'Curves' you can brighten the star data and keep the background sky dark at the same time. As soon as you click on it, you will be show a graph with a straight line. Using graph co-ordinates go along three and up three. Click on the line and move it up roughly to the point shown. The stars will now brighten and more will appear. The downside is the sky will brighten too. The next step will cure it.

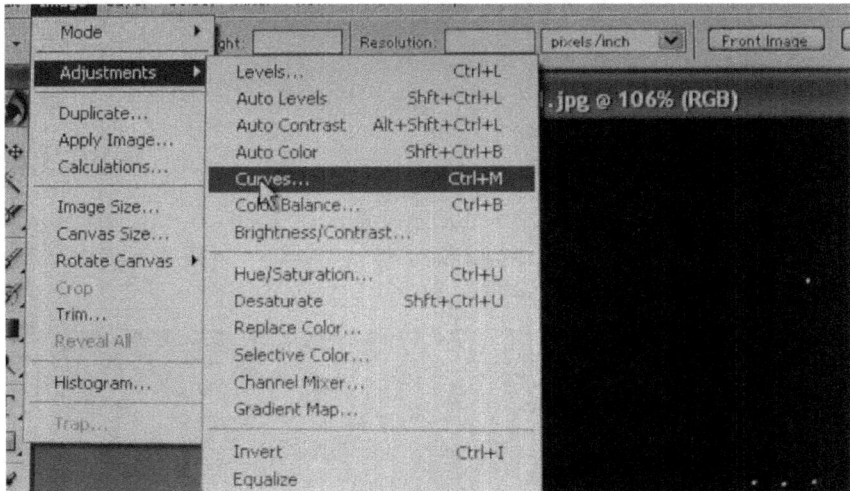

The rest of the line is now a curve, grab the line at around one across and two up and drag it down roughly to the point shown. This will now darken the background sky without dimming the stars. The best results will be from experimenting.

Simply study the graph; the two sides show dark to light. Move the dark part of the straight line that is shown to you a little toward the bottom to make that part of the picture darker still, and the lighter part of the line upward to brighten the stars further. More information that was hidden is now shown instead and the background sky darkens to reduce light pollution etc.

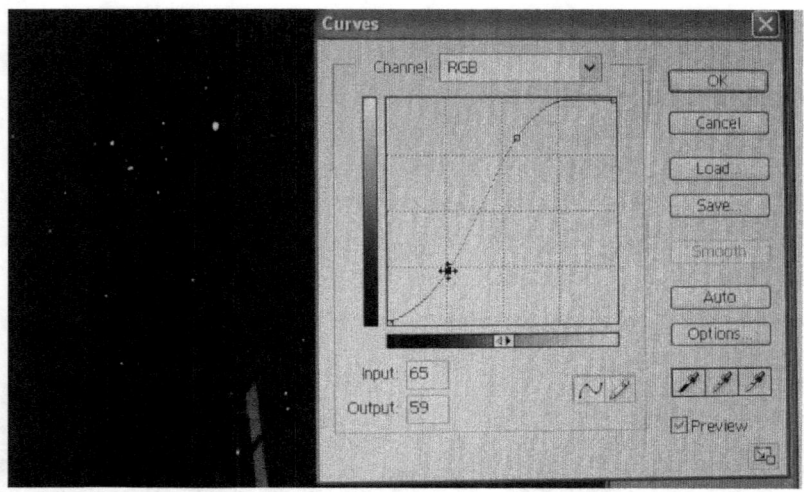

Save the picture with a new name and keep the original. As time passes, you will learn extra skills and may wish to alter the original image differently at a later stage.

The green glow has now gone and the number of stars has trebled. A satellite trail may have been recorded perfectly but seem faint on your original image. These two techniques alone will save your picture and motivate you a little more perhaps in achieving even better results in the future.

Try using the contrast option too but not too high a setting. A faint satellite image may be best displayed if the image was turned into a Negative or even Black & White.

Dozens of other tips can be used to enhance your images and will be dealt with in future editions of the website.

One more example of before / after comparison...ISS is seen fading into the Earth's shadow. Photo by a long-term friend; Tony Rickwood of Ullapool, Scotland.

Chapter 18   Favourite Targets

There are numerous targets in the sky for beginners as well as experienced observers. I have chosen a small number that are interesting from a human standpoint, observational interest and historical.

International Space Station (ISS): By far the brightest and most well-known satellite of them all is the International Space Station. When this fantastic piece of engineering passes over your hometown, it really brings home that this technology is for all of us to wonder at and be proud of. It should remain operational and seen in orbit until at least 2025. The Sun is the brightest object in our sky; the Moon is next; this is in third place.

NASA image November 2020

Spotting and imaging this satellite is an excellent way for anyone get interested in this hobby. On board will be human beings flying overhead at 7km per second. It can be seen twice an evening or morning.
Do not forget to wave as they go by.

You may also be lucky enough to witness a docking or undocking event. A second fainter object will then be seen a few degrees from it.

As of November 2014, Russia is currently looking at pulling out of the ISS project in 2020 and building their own space station once again. A political decision rather than a scientific one regarding the troubles in Ukraine and the US / European / Russian relationship probably cause this initiative.

ISS is by far the easiest satellite to image. This is a 10-second exposure; Jupiter is the bright star to the lower left of the trail. The tree provides scale.

Mars Ship? It is possible to transform ISS into a ship capable of ferrying a crew to Mars. It has all the facilities required – power, crew quarters, observation deck, food storage etc. The only extra hardware required would be two Mars decent vehicles to be docked on and a re-startable engine. Its current velocity is already 70% that is required for a journey to Mars (Several astronauts have agreed to it in principle including John Blaha – visited ISS several times via Shuttle and Soyuz).

Envisat: An Earth observation satellite. Its objective was to service the continuity of European Remote-Sensing Satellite missions, providing additional observational parameters to improve environmental studies.

In working towards the objectives of the mission, numerous scientific disciplines use the data from the different sensors on the satellite. They study atmospheric chemistry, ozone depletion, biological oceanography, ocean temperature and colour, wind, waves, hydrology (humidity, floods), agriculture natural hazards, digital elevation modelling, maritime traffic, atmospheric pollution, cartography and also of snow and ice.

Envisat cost £2billion including operational, development and launch costs. The mission is now replaced by the Sentinel series of satellites.

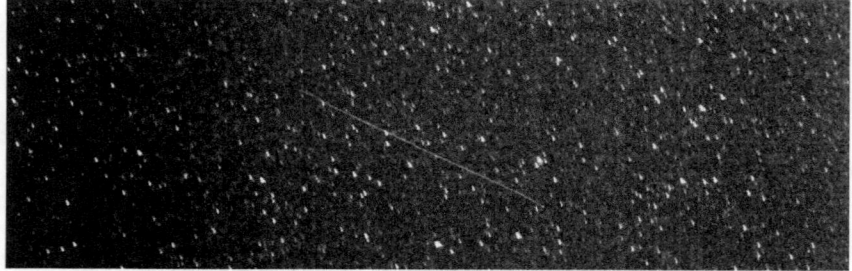

Envisat Trail over Huntsville, Alabama.

Size of target 26mt (85ft) × 10mt (33ft) × 5mt (16ft). Its size and low orbit defines a constant brightness of between Mag 2.5 & Mag 3.5; bright enough to be seen from most towns. Its orbit is 100 minutes long, 774km high almost circular at 98° and repeats the ground track every 35 days. So this is visible most of the year from all over the world. In 2012, contact was lost and is drifting in orbit, but it will remain visible for decades to come until its orbit naturally decays.

ESA image.

Tiangong 2: China's first space station was Tiangong 1; serving as both a manned laboratory and an experimental test bed to demonstrate orbital rendezvous and docking capabilities. It was launched unmanned aboard a Long March 2F/G rocket on 29 September 2011. This space station ended its life in April 2018 and was replaced by Tiangong 2.

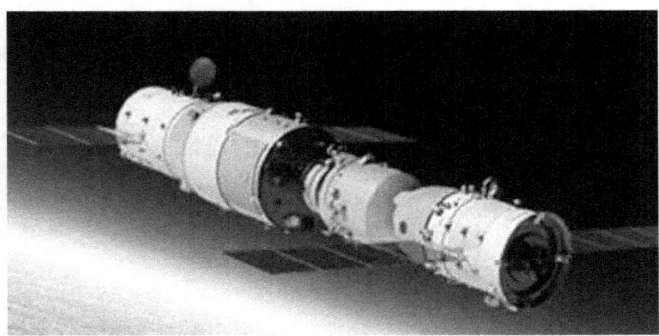

This is a very bright satellite, often passing over the southern parts of the UK but low down, as the inclination is only 42.8°; the south coast of England is 51°. It can only reach a maximum of 20° above the southern horizon.

119

Explorer 7: Launched on 13 October 1959 by a Juno II rocket from what was called then Cape Canaveral into an orbit of 573 km by 1073 km and inclination of 50°. It was designed to measure solar x-rays and other energetic particles and heavy primary cosmic rays. Secondary objectives included collecting data on micrometeoroid penetration and molecular sputtering and studying the Earth-atmosphere heat balance.

The satellite mass is 41.5kg, 75cm x 75cm. Powered by solar cells it also carried 15 nickel cadmium batteries around its middle to even out the mass to stabilise any rotation. It transmitted data continuously through to February 1961 and went dead 24 August 1961. The orbital inclination is at 50° therefore visible out to 55° or so north and south of the equator. This includes the UK and is one of the oldest satellites that can be seen. However, due to its small size, it is not very bright, maximum of Mag 5.5, only just visible to the naked eye in perfect conditions. Easy binocular object when it is at lowest point of its orbit – now at 520 km due to atmospheric drag. Search Heavens-Above for its position.

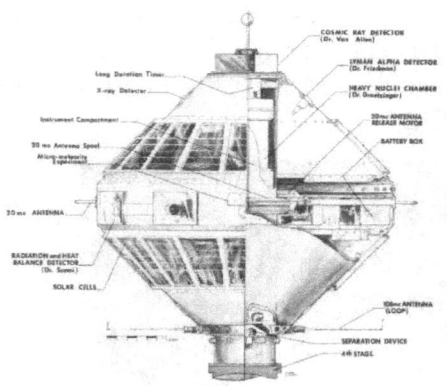

Prospero: The only satellite built and launched by the UK. The object is only around 1metre across and its minimum altitude is 527 km. This is only visible in binoculars and telescopes, but is very rewarding for folk in the UK. This was launched in 1971 on a Black Arrow rocket at Woomera, Australia. The satellite is no longer transmitting. The orbit is expected to decay by around 2070.

A full size model of Prospero at Alum Bay on the Isle of Wight. Walk to the top of the hill facing the Needles Lighthouse and pass round the corner. A free museum houses many artefacts from this era.

A full-size replica of the Black Arrow is on display at Sandown Airport, Isle of Wight. During the summer months, it stands upright. The author has produced a YouTube film of this and is shown in the Science Museum, London. Check out our website - page 'British Space Program.'

Then and now 1969 / 2020 from a drone. The UK launch test facility at Alum Bay, Isle of Wight UK is open to visitors and completely free. Lower two photos by the author.

The 'hold bolts' that restricted the rocket from launch during tests.

Prospero - Visible Passes

Search period start: 17 May 2014 00:00
Search period end: 27 May 2014 00:00
Orbit: 527 x 1313 km, 82.0° (Epoch: 16 May)

Passes to include: ⦿ visible only ○ all

Click on the date to get a star chart and other pass details.

Date	Brightness (mag)	Start			Highest point			End			Pass type
		Time	Alt.	Az.	Time	Alt.	Az.	Time	Alt.	Az.	
17 May	8.4	22:06:12	10°	SW	22:12:24	32°	WNW	22:17:41	10°	N	visible
18 May	8.8	22:15:53	10°	WSW	22:21:25	25°	WNW	22:26:13	10°	N	visible
19 May	9.1	22:25:47	10°	WSW	22:30:28	20°	WNW	22:34:37	10°	NNW	visible
20 May	9.3	22:35:58	10°	W	22:39:33	15°	WNW	22:42:50	10°	NNW	visible
21 May	9.5	22:46:44	10°	WNW	22:48:41	11°	NW	22:50:31	10°	NW	visible
25 May	9.2	21:34:54	10°	WSW	21:39:06	19°	WNW	21:42:48	10°	NNW	visible
26 May	9.3	21:45:03	10°	W	21:48:07	14°	WNW	21:50:55	10°	NNW	visible

Use the 'Search Database' button on Heavens Above website to look for a specific satellite. As I became interested in satellite spotting, this was my first major ambition; to see and record Prospero passing over Britain.

Check out the 'British Space' page on the website.
The first satellite on this clip is Cosmos 1244 and Prospero is seen half way through the clip as the fainter object travelling in the opposite direction. The e-book version of this publication will take you straight to such clips.

This Black Arrow replica at Sandown Airport is probably the most photogenic model I have ever witnessed.

CryoSat 2: CryoSat 1 was a launch failure and fell into the ocean. However, the second was successfully launched in 2010 and partly built in the UK by Astrium. Duncan Wingham of University College, London, made the original proposal for the project. For this mission that includes 3D mapping of ice, extremely accurate satellite positions are required. This is achieved by a laser retro-reflector on board and a radio DORIS system that works by radio parallax from the ground.

After three years of construction at Stevenage, Hertfordshire, final testing took place in Germany and launched from Kazakhstan. It is under contract from the European Space Agency to monitor the changing conditions at the Earth's poles. The orbit you may have guessed is polar and passes over Britain every day. Its average orbital height is 710 km and is only 4mt x 2.3mt in size on the downward facing axis. This object is not very bright; around Mag 4, but is largely a British built satellite and gathers environmental data; the ultimate purpose of 21st Century space technology.

SeaSat 1: Launched in 1978 from Vandenberg Air Force Base, California. It is in an almost circular polar orbit around 745 km high. This is a relatively slow crossing satellite; about 70% of the speed that ISS takes to cross. Its brightness is unusually high due to the solar panel surface area at between Mag 1.9 and 3. So the combined points make it a relatively easy target to photograph or capture as a moving image from any country.

Spy Satellites: Early spy satellites such as the US Corona series used to take photographs on black and white or Infra-red film Then it released a capsule with the exposed film on board These *'Buckets'* were picked up by a military team. They were designed to be retrieved during the parachute stage of recovery. They could float, survive desert or artic conditions, and transmit a location beacon on a secret frequency. The movie *'Ice Station Zebra'* is weakly based on a true story. The USA & the Soviet Union both launched such machines.

Modern spy satellites use digital cameras of course and have superseded the Buckets. They can now have an operational lifespan of a decade or so rather than weeks. The easiest to find is the US Lacrosse series; relatively bright and in a high inclination so most of the world can view them and most of the world is monitored from them too.

The cameras are always being superseded by new technologies developed down here so such satellites are upgraded often.

Such satellites can fly in formation in twos or threes as a triangle around 10 degrees apart.

One story that has leaked out regarding an early Lacrosse satellite goes back to the 1990's. Colonel Gadaffi was hiding as usual somewhere in the Libyan Desert. The CIA wanted to determine his exact position and employed the best spy satellite they had. Gadaffi employed a British (sorry about that) satellite spotting expert to predict when each US spy camera passed over his camp and blew a whistle minutes before each pass. Everybody dived into their tents.

Gadaffi became tired of this game and purchased some Colonel Gadaffi masks. They were given out to everyone on the camp and next the next time the whistle blew, everyone donned the masks, stood still outside and faced the sky. Gadaffi himself wore a mask as an infrared image combined with optical may have shown whom the real one was. The CIA realised that they had been rumbled and gave up.

Hubble Space Telescope (HST): The orbit of the HST is at an inclination of 28.5°. Taking into account the angle of sight from Earth, it can only be witnessed between around 38° north or south of the equator. Southern UK is at 51°; no chance here. The reason for this placement was simply economy. The closer the orbit is to the equator, the more advantage can be taken of the Earth's rotational spin; less fuel. The closest equatorial orbit achievable from the Kennedy Space Center is 28.5°. Observers in the southern US will see it.

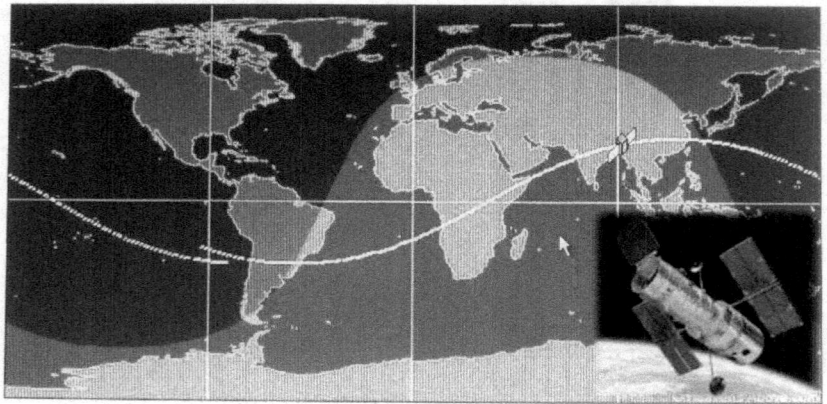

(This has an extra special meaning to me as I saw it launched by the Discovery space shuttle in April 1990).

The satellite prediction maps include the whole sky as if viewed by a fish-eye lens. The satellite path will be shown as a curve.

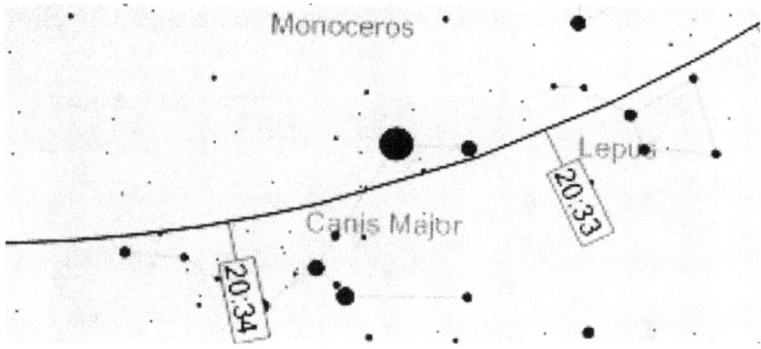

The exposure below is of the Hubble Space Telescope from Huntsville, Alabama in 2014. It was just one degree lower than the predicted path. This can be a bright target due to its size and the orientation of its solar panels.

The bright star is Sirius in Canis Major to the lower right of Orion. You can make out the whole of Lepus the Hare to the right of the image.

ICESAT 2: This satellite launched on 15th September 2018 from Vandenberg Air Force Base, California on a powerful Delta II rocket. Its function is to map accurately the thickness of sea and land ice at the poles, glaciers and mountain ranges to within 4mm of accuracy; a measurement totally impractical and financially unfeasible to gain from the ground. A complex on-board system of laser-based radar (Lidar) allows this data to be accumulated and build up a global picture of ice trends connected with climate change.

This is a bright satellite that can be seen the world over due to its polar orbit. The solar panel can produce flares that can brighten its appearance from Magnitude +3 to -1. Search Heavens-Above for timings from your local area. It is well worth a look.

This trail of ICESAT 2 near the Lagoon Nebula was taken from Arizona in April 2019 by the author.

Starlink Constellations: A constellation of satellites constructed by the SpaceX Company to provide low cost internet to the globe. When complete, it will consist of thousands of mass-produced small satellites working with ground transceivers. SpaceX also plans to sell some of the satellites for military, scientific or exploratory purposes.

Starlink constellation phase 1 will consist of 1,584 satellites at 550 km altitude. As of January 2020, SpaceX has deployed 182 satellites. They plan 60 per launch, as often as every few weeks after late 2019. In total, nearly 12,000 satellites will be deployed by the mid-2020s, with a possible extension to 42,000. The initial 12,000 satellites will be in three orbital shells: first approximately 1,600 in a 550-km altitude shell, then 2,800 satellites at 1,150 km and approximately 7,500 satellites at 340 km (210 mi). Commercial operation for universal internet coverage begins in late 2020.

Concerns have been raised about the long-term danger of space junk resulting from placing thousands of satellites in orbits above 1,000 kilometres (620 mi) and a possible impact on astronomy, although SpaceX is reportedly attempting to solve the issue. The frequencies transmitted are not of interest to radio astronomers, but they can be a nuisance for astrophotographers, as these will often pass in front or near an object being imaged.

SpaceX estimated the total cost of the decade-long project to design, build and deploy the constellation in May 2018 to be about US $10 billion; it began with the first two prototype satellites launched in February 2018. A second set of test satellites and the first large deployment of a piece of the constellation occurred on 24 May 2019 when the first 60 satellites were launched. A dedicated page is on the website.

For the Starlink constellations, use Heavens-above, ensure you choose your observing location button, click update, then click on List of Brightest Satellites and look for a grouping as shown below. They are all bright as the orbit is comparatively low and appear seconds from each other. Pick opportunities where they are at least 40° above the horizon. In the example given, they are all above 80° - virtually overhead.

Sat name	Brightness				Time	Height
STARLINK-1284	1.8	21:04:44	10°	WSW	21:07:51	81°
STARLINK-1265	1.8	21:05:11	10°	WSW	21:08:18	82°
STARLINK-1258	1.8	21:05:38	10°	WSW	21:08:46	82°
Okean 3 Rocket	4.1	21:04:31	10°	NNW	21:08:58	62°
STARLINK-1019	3.3	21:05:18	10°	SSW	21:09:00	32°
STARLINK-1296	1.8	21:06:05	10°	WSW	21:09:12	82°
STARLINK-1292	1.8	21:06:32	10°	WSW	21:09:39	83°
STARLINK-64	2.2	21:06:25	10°	WSW	21:09:58	68°
STARLINK-1260	1.8	21:06:59	10°	WSW	21:10:07	83°
STARLINK-1274	1.8	21:07:26	10°	WSW	21:10:33	83°
Cosmos 2360 Rocket	3.1	21:05:28	10°	NNW	21:10:38	35°
Cosmos 1821 Rocket	4.1	21:04:14	10°	N	21:10:43	87°
STARLINK-1318	1.8	21:07:53	10°	WSW	21:11:00	84°
STARLINK-1289	1.8	21:08:19	10°	WSW	21:11:27	84°
STARLINK-43	2.6	21:08:09	10°	W	21:11:51	66°
STARLINK-1275	1.8	21:08:47	10°	WSW	21:11:55	85°
STARLINK-1306	2.2	21:08:26	10°	WSW	21:12:07	79°
Akebono	4.2	21:09:54	10°	S	21:12:07	60°
STARLINK-1303	1.8	21:09:14	10°	WSW	21:12:22	85°
BREEZE M DEB (TANK)	3.3	21:08:56	10°	WSW	21:12:27	66°
MetOp-A	4.4	21:07:21	10°	S	21:12:42	72°
STARLINK-1309	1.8	21:09:41	10°	WSW	21:12:49	85°
STARLINK-1312	1.8	21:10:07	10°	WSW	21:13:15	86°
STARLINK-1282	1.8	21:10:35	10°	WSW	21:13:43	86°

Passing Probes & Junk: If anyone is looking for a real challenge, every so often a deep space probe or an old spent booster passes the Earth. Probes to other targets around the solar system usually pass around 1000-5000km from us and may be on their way to a much more distant target and use the Earth's gravity to accelerate the craft further.

On April 10th 2020, 2.40am, the European probe Bepi Columbo passed us on its way to Mercury. This flew by in just over an hour. At Magnitude 9.5, it was very faint, but not impossible to observe, many people did but I remained in bed. Full star-maps with times are announced in advance of such events.

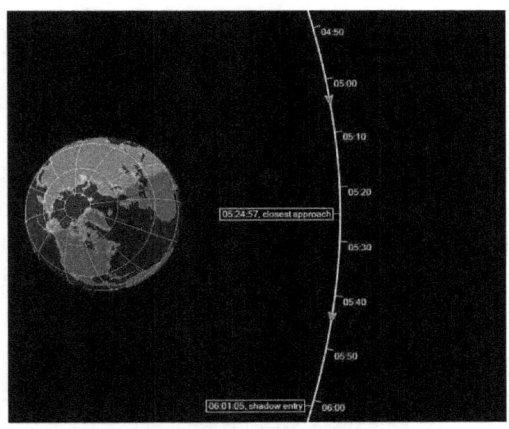

Other passing probes that have been witnessed include the Venus Express, Messenger while on its way to Mercury and Cassini on way to Saturn; that flyby was in August 1999.

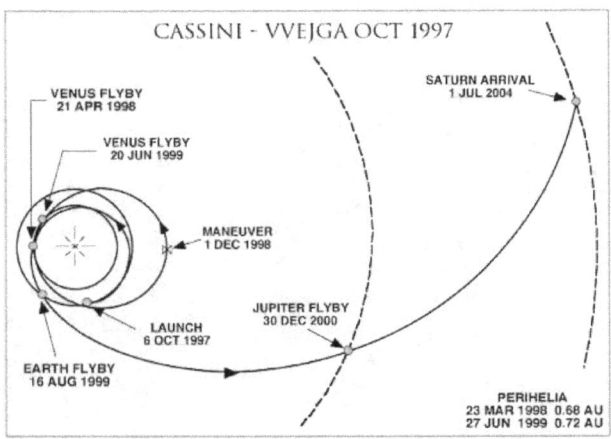

Historic Junk: In 2002, an object was detected very close to the Earth and was given the name of Asteroid J002E3. After closer examination, it was discovered to be made of aluminium and was a perfect cylinder around 10 metres long. It turned out to be the third stage of Apollo 12 that propelled the Command & Lunar Modules to the Moon in 1969; not aliens. It continued in a changed orbit around the Sun; the Earth's gravity had altered it. The defunct third stage was visibly seen crossing the night sky by amateur astronomers.

Another similar discovery was made September 2020. It was classed as an asteroid with the assigned name of 2020SO. As it's path was taking it close to Earth, NASA felt it necessary to research it deeper. Their Infrared Telescope Facility (IRTF) was turned toward the object and found it was made of stainless steel. After calculating its path backward in time, the conclusion was that it was the upper stage of the Centaur rocket that launched Surveyor 2 toward the Moon in September 1966. It remained in Earth / Moon vicinity until March 2021.

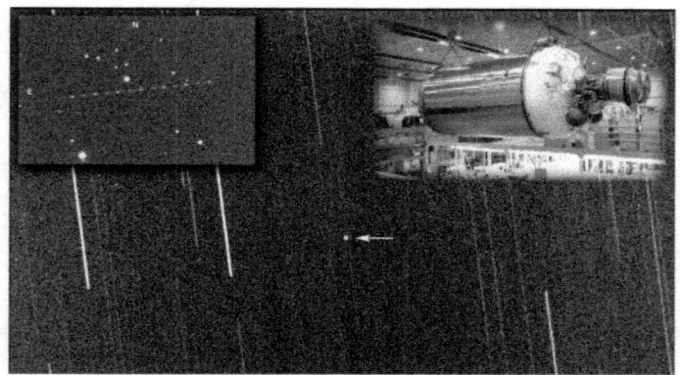

Blue Walker 3; a giant unfolding satellite that is the world's first to allow a direct 5G line between a satellite and a phone. Developed by AST SpaceMobile, it unfolds to reveal a flat antenna spanning 693 square feet. The prototype unit entered orbit in September 2022, but a series of other models called Blue Birds will soon be orbiting.

The Blue Walker 3's brightness reached an average of Magnitude +0.4, which was greater than some well-known stars such as Betelgeuse and Altair, putting it in the same league as the top 10 brightest 'stars' in the sky. As it catches the sun and reflects it downward, it creates a flare of light that outshines every star apart from the Sun.

I personally feel that this will become a favourite target for satellite observers in the near future.

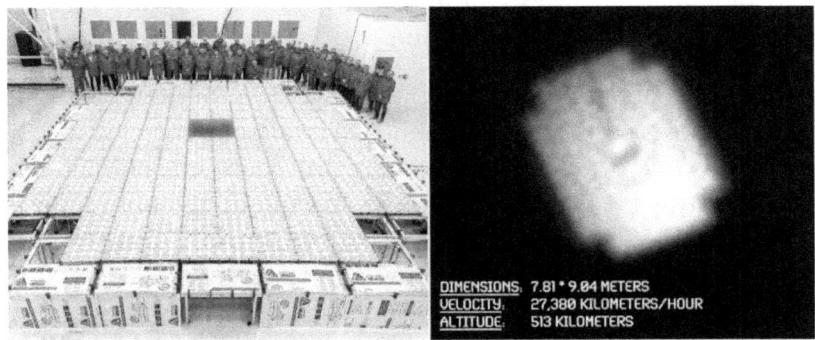

DIMENSIONS: 7.81 * 9.04 METERS
VELOCITY: 27,380 KILOMETERS/HOUR
ALTITUDE: 513 KILOMETERS

Credit; AST SpaceMobile.

Bright satellites are often viewed as an annoyance to amateur and professional astronomers alike. It can ruin images of distant stars and galaxies. However, as we saw in the chapter 'Why Can We See Satellites' for much of the night and most of the year, satellites are orbiting within the Earth's shadow. During such passes, no satellites are seen. During the winter months, satellites are only observed for the first and last 85 minutes or so of darkness. Plus, the closer the observer is to the Equator, the higher the shadow becomes throughout the year; less satellites are seen per night.

Chapter 19   Transiting Satellites

Every now and then, satellites, as well as aircraft and balloons can pass in front of the Sun and Moon. This is known as a Transit. The planets Mercury and Venus can Transit the Sun on rare occasions. As the Sun and Moon are just 0.5° across and most satellites travel at around 0.25° per second across the sky, photographing the event is somewhat tricky; but not impossible. The higher satellites in mid-Earth orbit or above take several more seconds to cross.

Several satellites a day will Transit the Sun or Moon and can be imaged even without using the Internet. All you need is one skill – patience. The best time of the year is when the Sun crosses the equatorial plane – 21 March & 21 September. This is when the geo-stationary satellites can transit the Sun as they are directly above the equator; so will the Sun.

When I was about 13-14yrs old, I imaged the Sun every single clear day when I was off school. I waited until mid / late afternoon when the Sun shone into my bedroom, pointed my telescope toward the Sun and projected the image onto a large white card. I could produce it up to 50cm across. I darkened the room with thick blankets but allowing the sunlight to travel into the telescope. I observed the sunspots change shape by the hour. Every now and then a plane would fly in front of the Sun and saw its silhouette complete with vapour trail.

On rare occasions, a small dot would take a few seconds to cross the Sun and I was baffled as to what it was. 'UFO' I thought straight away. After a little research I realised I was just seeing satellites. 'Ohhh shucks.'

Many objects in our skies are claimed to be Flying Saucers to an inexperienced observer. It is understandable, but we do need to be realistic about this subject and be very careful not to influence such reports by our desire to want them to be real.

My solar projection method in 1977. A thick blanket was pegged up to darken my bedroom and the telescope was pointed out the window toward the Sun. Tracking the moving Sun was made by turning the telescope controls.

With experience using satellite prediction websites, preparation can be made to see the larger satellites transit the Sun or Moon. But please never look through any instrument at the Sun unless you are using the appropriate filters please. Better still, project the image as shown.

The space shuttle Atlantis transiting the Sun on mission STS125 in 2009. Photo by Theirry Legault from Florida.

Orbiting satellites occasionally pass in front or very close to the Sun in the sky. The Sun is very noisy in the radio spectrum and often interferes with the satellite transmissions. These 'Sun-out' events with Microwave transmissions hardly suffer from this effect as the Sun is quiet in the microwave frequencies. All the TV satellites use microwaves for this very reason.

In the Northern Hemisphere, solar transit is usually in early March and October. In the Southern Hemisphere, solar transit is usually in early September and April. The time of day varies mainly with the longitude of the satellite.

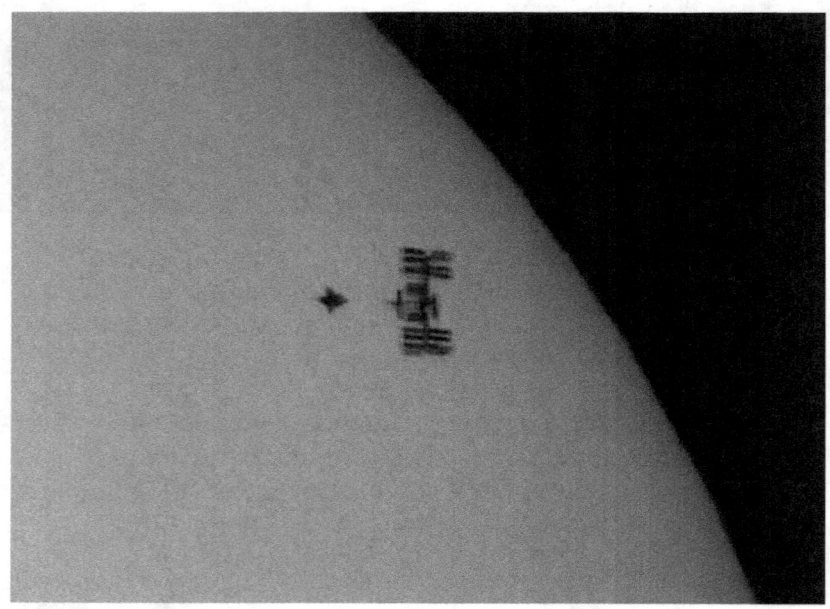

Another incredible shot by Theirry Legault of ISS and the space shuttle Atlantis in 2010 from Spain this time.

UFOs: It is understandable that some people really do believe that these dots passing in front of the Moon or Sun are flying saucers. We know they are indeed satellites; the larger machines can reveal detail, the smaller or more distant will not. There are many clips on YouTube by people who are totally convinced they have imaged aliens flying over the Moon when in reality they have imaged a satellite perhaps 10,000km up passing in front of the Moon that is 400,000km away. The higher the altitude, the slower it passes. Just type in 'UFO Moon' in the www.youtube.com search box and choose your own clips to laugh at. Politely explain to them what they have really imaged. Examples are on our website.

The 'UFO's do not jump around as some claim either. It is being recorded by a digital chip where the pixels are arranged in rows. As the satellite's tiny image crosses each row, it just *seems* to jump as it passes from one row of pixels to the next.

An image of ISS transiting the Moon.

Ed Morana specialises in satellite transits. Try his internet page for other links on how to observe them yourself. Such information changes daily so this book cannot keep up. The e-book makes these links more accessible / or use the website and look up the page on 'Transiting Satellites.'

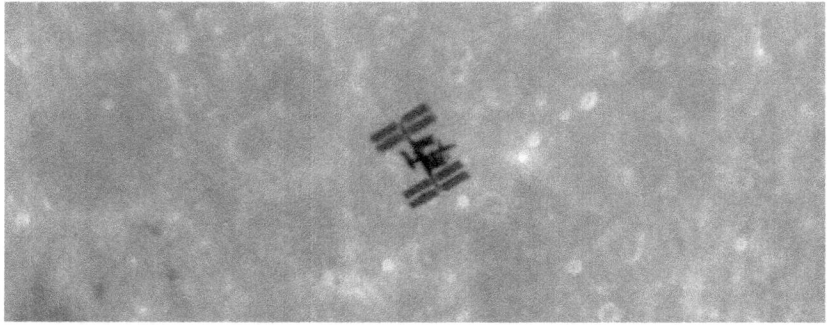

Chapter 20   Re-entry

Even though satellites are generally orbiting in the vacuum of space, there is still a trace of atmosphere buffeting the machine. There may be just a few air molecules a second hitting the satellite that may have a mass of several tons. But this weak force can have an effect on the orbit; month by month it will lose speed and drop gradually toward the earth. This in turn will increase the number of air molecules encountered per second and accelerate the process.

During a satellite's useful life, on board engines will fire to boost its altitude back up toward its parking orbit. After a decade or so, either the fuel will run out or the satellite has now served its purpose. The orbit can now be allowed to decay and finally enter the earth's atmosphere at just under 7 km per second. The final orbit will only be around 65-70 km above the earth.

The very first confirmed crash landing of a satellite was a piece of Sputnik IV that landed in the main street of Manitowok, Wisconsin, USA in 1958. It was sent back to Russia and never received any thanks. A metal ring in the street marks the historic impact point to this day.

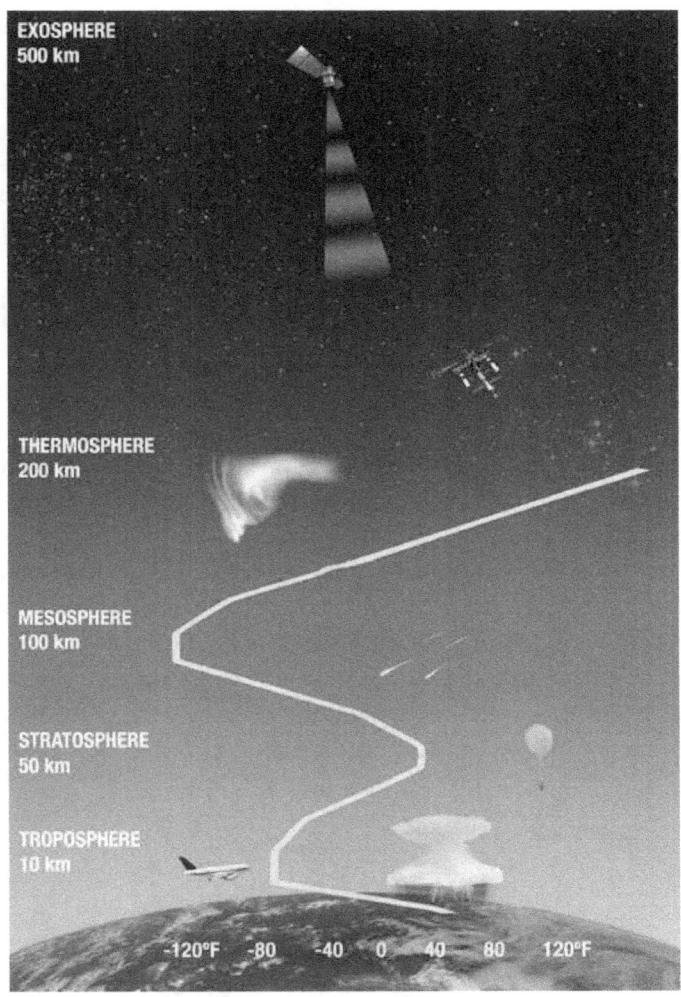

Even though satellites are hundreds of kilometres up well into the Exosphere, these still is a few molecules of air per cubic centimetre. This is enough to slow a satellite down over time. This will reduce its velocity and bring it lower still, more are molecules are encountered per second so slowing it further.

Satellite re-entries can be quite spectacular and viewed by large numbers of people, as the craft will often be seen over a few thousand km before it finally succumbs. The first example was Sputnik 2 and was witnessed by many people during its descent on 14 April 1958. For ten minutes it travelled from New York to

139

the Amazon, it descended from 130 km to around 60 km, leaving a trail of `sparks' some 100 km long, the satellite itself produced a multitude of brilliant colours.

To reduce the increasing hazard of defunct satellites, a global policy now exists to reserve some remaining fuel to deliberately slow the satellite down at this last stage and force it to burn up within a few days or weeks. Such events are a little more predictable than the random re-entry. This is known as the de-orbit procedure.

Skylab: Perhaps the most infamous of all the re-entry events was Skylab. This 77-ton spacecraft was launched in 1973 and was operated by 3 teams of US astronauts. After its final mission, it was left to drift in orbit and began to come down. By 1975, plans were underway for a space shuttle mission to attach a booster to it and push it back up to a safe parking orbit; $750,000 was spent on this investigation. It would have been carried out on the shuttle's fifth mission sometime in 1980. There was even some talk of refurbishing the $1.1 billion space station for further missions. As time passed, it became obvious the shuttle would not be ready in time, the first launch date slipped into 1981.

The shuttle program was plagued with technical delays and the Sun became more active. This expanded the upper atmosphere and increased drag on all low satellites A new predicted re-entry date was sometime around May 1979 rather than 1983 as predicted by NASA. This was a result from a study carried out by the Royal Aeronautical Establishment (RAE) in England. Satellite spotters (myself included) timed and plotted the orbit and submitted them to the RAE.

The North American Air Defence Command (NORAD) located the workshop by radar, aimed a radio signal at it and received a response. For two minutes, Skylab reported on the condition of its systems. It was rotating at about 10 revolutions per hour and when its solar panels turned sideways to the sunlight the radio transmissions ceased. Amateur satellite observers confirmed this spin by measuring the changing brightness.

The first thing the engineers needed to do was to charge the batteries and since they could transmit commands only briefly once during each orbital pass, this would take time. Within a week, however, they had charged two batteries, determined the workshop's orientation. The onboard computer was used to help control the spacecraft via its Reaction Wheels. The aim was to decrease drag by rotating the panels parallel to the Earth and keep it in orbit longer for a shuttle mission, but too much altitude was lost already.

During its last moments, NORAD managed to rotate the spacecraft in a sideways orientation to increase drag and force it to comedown somewhere over the Pacific Ocean. As such, an experiment had never been tried before and data on the upper atmosphere at that time was limited, it was incredibly difficult to control a re-entry point. It was hoped for it to come down over the Pacific and most of it did. NORAD computed that impact finally occurred at 12:37 pm Washington time on 11 July 1979. Shortly before 1pm, NASA at their Washington HQ received word that the area southeast of Perth, Australia, had indeed been showered with some pieces.

Spectacular visual effects were reported and many residents heard sonic booms and whirring noises as the chunks passed overhead in the early morning darkness. The majority did indeed come down over the Pacific. This was almost certainly down to good fortune on the day rather than planning. Recovered debris is now on display where these components fell. Some has been moved to Huntsville, Alabama and is on display at the Marshall Spaceflight Center.

Esperance Municipal Museum in Australia; several recovered pieces of Skylab are on display.

A fuel tank on display at the Marshall Spaceflight Center, Huntsville, Alabama.
(Image by the author 2010)

Some vehicles are planned to re-enter are under full control by a space agency or company. Such events are far more predictable and remarkable sightings are made. Unfortunately for many of us they are normally made over an ocean. The next image is of a service vehicle for ISS called the Jules Verne ATV module burning up over the Pacific Ocean. *(Image supplied by ESA).*

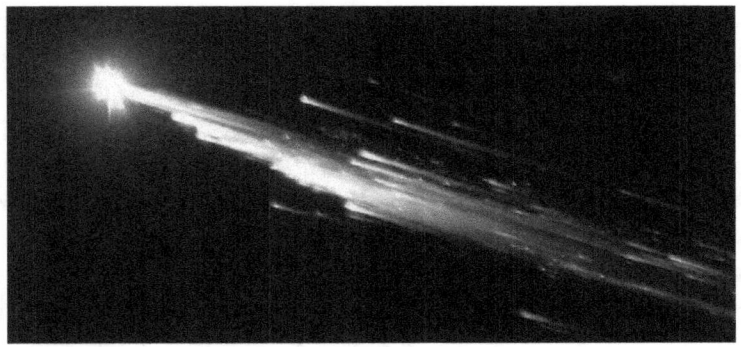

On 25th July 2024, my wife and I were visiting friends on Oak grove, California. On our very first night, we witnessed a re-entry event of an unknown satellite. There was no sound, just an incredible display of light with a dense smoke trail behind it. We recorded the sighting on video and took a screenshot below. A full YouTube movie is linked on the re-entry' page on the Satellite Spotting section of www.outerspacebooks.com

Satellites that are naturally falling from orbit are announced on satellite prediction websites such as Heavens-Above. Announcements are made to observers to look out for such events if you happen to be under a predicted fall. I have only ever witnessed two re-entry events, the first from a stargazing camp in Sussex, England. I don't know what satellite it was but it lit the sky for around 30 seconds and produced beautiful green and blue colours. The different melting materials produce these during break-up. Several re-entry clips are shown on our website.

Cosmos 954: On 24 January 1978, NORAD tracks a fireball streaking across the skies over the Canadian Northwest. Cosmos 954, a Soviet nuclear powered satellite, crashed near Great Slave Lake, scattering radioactive waste across a 124,000 square kilometres of tundra through Alberta and Saskatchewan. There urgent questions for Canada's Prime Minister Trudeau: Why wasn't there more warning? Were the Americans and Russians holding back information? In addition, who will clean up the mess? Who is going to pay for it? The Soviets denied it even existed.

A radioactive portion of the craft fell near a trapper's camp. A canoeist travelling through found it, looked at the unusual item and then left it alone. A massive search was begun to locate the rest of the debris, it was made up of hundreds of troops from Canada, Britain and the US. The search lasted until October. Tight security was present and no civilians were permitted. The canoeist and his radioactive find were taken back to Yellowknife; the canoeist was found to be in good health and the reactor pieces were impounded.

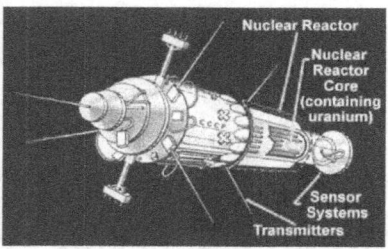

After the clean-up, the Canadian Government sent a $15 million bill to the Soviets. They paid less than half of this amount and agreed not to take back the spacecraft. The Canadians were content with the amount they received but were happier still that the Soviets had finally acknowledged the spacecraft's existence.

This incident raised international policy questions. There was a request from the US to prohibit satellites containing radioactive material from orbiting the earth. This was followed by similar calls from Canada and Europe. In November 1978, the UN authorised its Committee on the Peaceful Uses of Outer Space to ban nuclear-powered satellites in Earth orbit.

MIR Space Station: Until the construction of the International Space Station, (ISS) MIR was the largest structure ever built in space. The inevitable end was bound to be a fiery one that required careful planning or somebody could conceivably get hurt.

The de-orbit of Mir was a controlled atmospheric re-entry of the modular Russian space station. Major components ranged from about 5 to 15 years in age. With MIR at the bottom of the ocean, it allowed Russia to channel its efforts on the new ISS project.

It was carried out in three stages. The first was waiting for atmospheric drag to decay the orbit an average of 220 km. This began with the docking of Progress M1-5. The second was the transfer of the station into a 165 × 220 km orbit. This was achieved with two burns of the Progress M1-5's control engines on 23 March 2001. After a two-orbit pause, the third and final stage of Mir's de-orbit began with the firing of the Progress engines at 05:08 UTC, lasting a little over 22 minutes. The final re-entry at the altitude of 100 km occurred at 05:44 UTC on 23 March 2001 and broke up over several minutes.

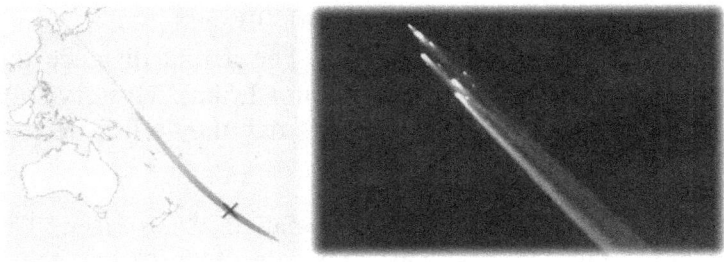

Break up of the MIR Space Station 23 March 2001 over the Pacific Ocean.

Antares Rocket: This map shows the re-entry path of an Antares rocket body that ended its fiery journey in the Indian Ocean as indicated on the yellow track. The blue line represents the predicted re-entry point. As you can gather, it continued for another 40 minutes or so half way around the world. This is not an exact science, so amateur satellite spotters do play an important role in this field.

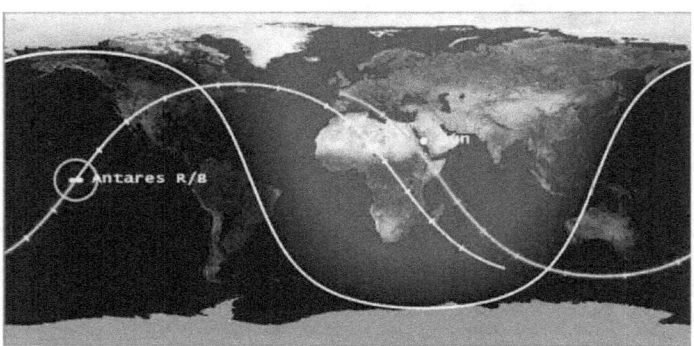

A useful link to re-entry events... **www.aerospace.org/cords/**

UFOs: To the inexperienced observer, these re-entry events are often reported as UFO's. On the 1st August 1974, dozens of witnesses including two prominent UFO 'experts' David & Barry Biedny made a sighting over Venezuela. They estimated their UFO to be around 1.5km away. Later analysis showed that this was a part of a rocket that was launched from Baikonur, Kazakhstan on 29 July. It was a Proton vehicle third stage and was left in a decaying orbit of 51°, 182 x 198km. It was 6.5m long, 4.15m in diameter and a mass of 3,500kg. It was assigned the international designation 1974-060B and a US Strategic Command catalogue number 7393. The actual distance to the sparkling debris was around 300km not 1.5km. Mystery solved! But David & Barry still insist to this day they witnessed a real UFO while ignoring the data.

The case is included in the TV series called *'Alien Files.'* This is the most inaccurate investigative reporting I have ever seen. It is an interesting subject, but scientific methods should be applied to gain more support rather than broadcasting biased guesses.

Point Nemo: An area well away from land has been chosen as a target for many satellites to end up. This is nicknamed as the 'Satellite Graveyard' or Space Cemetery.' Its official tittle is *Point Nemo* - named for the fictional submarine captain in Jules Verne's 20,000 Leagues Under the Sea. This is the furthest point from any landmass on earth, and 4km under the sea.

When their outer space journeys come to an end, old satellites, rocket parts and space stations are sent to this remote spot in the Pacific Ocean to rest on the dark seabed forever. The technical name for this stretch of water is the *"Ocean point of inaccessibility"* because it lies about 2,700km from any land. The ISS is expected to end up here around the year 2030.

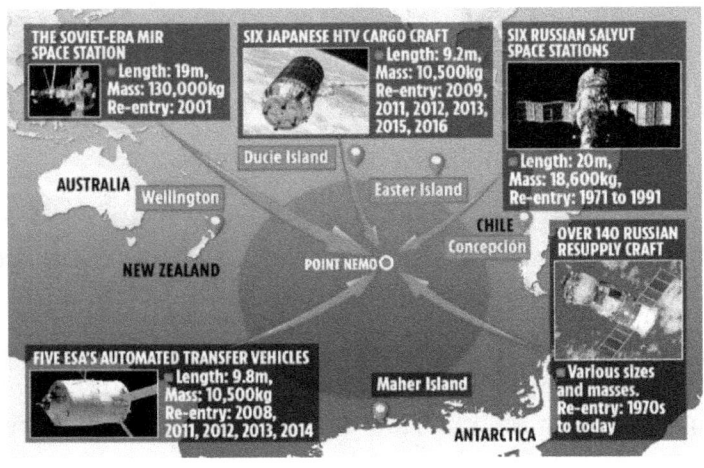

The furthest point from any land mass has been chosen as a target for semi controlled re-entry target for satellites from all nations... Point Nemo. Credit – The Sun newspaper.

Aeolus Satellite; a UK-built satellite mission launched by the European Space Agency (ESA) in 2018 as part of its Earth Explorer programme, reached the end of its active life in orbit and faced a final journey back to Earth. A friend of mine, Dr David Rees, invented a laser-based system for determining wind velocities at differing heights within the Earth's atmosphere. Previously, cloud movement observations alone gave information on wind velocity and direction; an indirect measurement. His gadget was chosen above a similar NASA

147

proposal. David built the final instrument to the required specifications and it was inserted within the satellite at Stevenage, UK. The original experimental mission target was for 1 year of operations; it lasted in full operation for 3 years and coasted for 2 years.

Credit Airbus Space & Defence, UK

Aeolis was constructed before any de-orbit regulations were established, this weather satellite was originally designed to naturally re-enter the Earth's atmosphere like most others.

By 2023, almost of its fuel reserves had been used up. ESA and its partners have embarked on a first of its kind task to carefully control Aeolus' re-entry. This complex series of manoeuvres have been planned and analysed over several months to ensure the spacecraft's-controlled descent as much as possible.

At ESA's Space Operations Centre in Germany, commands were transmitted to use the last remaining fuel to steer Aeolus during its return to Earth. The 1,360kg satellite had its velocity decreased to such a degree that gravity took hold and brought it safely down in the Atlantic Ocean. Radar tracking confirmed the re-entry corridor was as predicted. The vast majority of its mass vaporised, a small amount probably did make it to the ocean surface.

Chapter 21   Further Reading / links

http://www.satobs.org/satintro.html
A starter satellite prediction website.

www.heavens.above.com
The finest satellite prediction website in the world.

www.moonconnection.com
Perfect for Moon phase predictions. During around full Moon only the brightest satellites may be seen.

www.satlist.nl/index.html
A listing of every satellite launched since 1957.

www.satsig.net/sslist.htm
Listing of all the Geo-stationary satellites along with the Longitude data. 0 degrees in line with Greenwich, visible from that half of the world. 180 degrees, in line with the centre of the Pacific.

www.tbs-satellite.com
Fabulous site for giving up-to-date technical details on most satellites launched.

www.satview.org/
A tracking website for the most popular satellites such as ISS

www.heavenscape.com
Satellite Tracking Software

www.biotrack.co.uk
Animal Satellite Tracking Company

www.nro.gov
US National Reconnaissance Office

Book; Observing Earth Satellites by Desmond King Cole
Book; A vertical Empire – The history of the UK rocket
program 1950-1971 by C. N. Hill.

There is a smart phone 'App' that can track some satellites. This is in the early stage of development; they have had reports of inaccurate positioning. The Heavens-Above version has an experimental App for Android users. In time these will improve no doubt.

Satellites are changing the way we view the world and is an important tool to save the world.

Readers; if you have suggestions or photographs for future editions of this book please do submit via
www.outerspacebooks.com

Many thanks… Pete Bassett

Chapter 22   Other Books by the Author

**For the very latest listing of all books and versions,
use www.outerspacebooks.com**

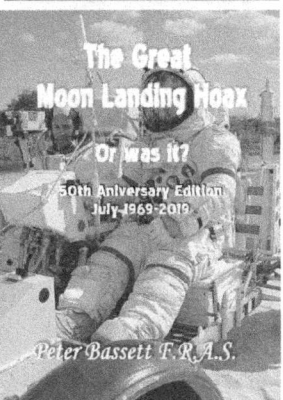

There are B&W, Colour and e-book versions.
All book sales support four charities; Cancer Research UK,
Kent Air Ambulance, Smile Malawi Orphanage in Africa,
British Hedgehog Preservation Society.